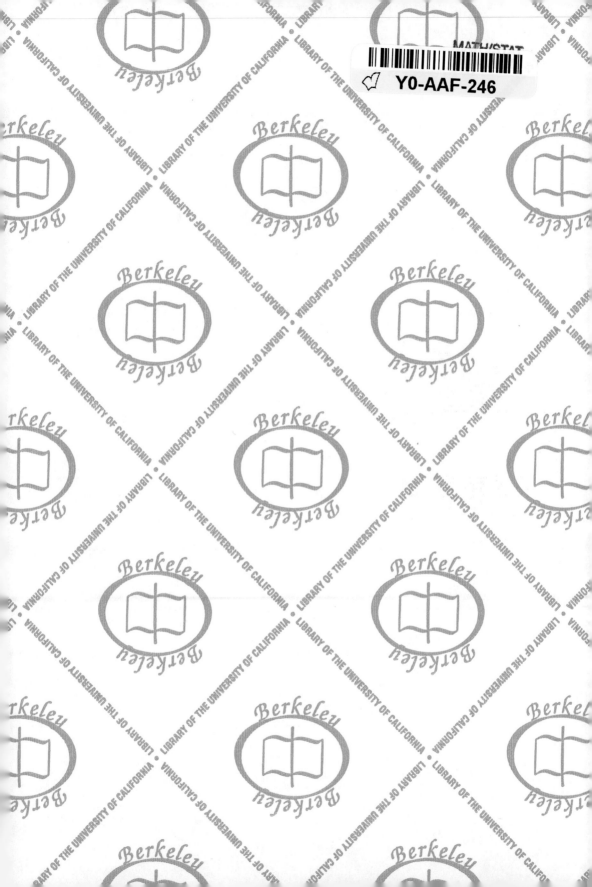

# Frontiers in Mathematics

Radu Zaharopol

# Invariant Probabilities of Markov-Feller Operators and Their Supports

Birkhäuser Verlag
Basel • Boston • Berlin

Author's address:

Radu Zaharopol
Mathematical Reviews
416 Fourth Street
Ann Arbor, MI 48104
USA

e-mail: rxz@ams.org

2000 Mathematical Subject Classification: Primary 37A30; Secondary 28D05, 47A35, 47B65, 60J05

A CIP catalogue record for this book is available from the
Library of Congress, Washington D.C., USA

Bibliographic information published by Die Deutsche Bibliothek
Die Deutsche Bibliothek lists this publication in the Deutsche National-
bibliografie; detailed bibliographic data is available in the Internet at
http://dnb.ddb.de.

ISBN 3-7643-7134-X Birkhäuser Verlag, Basel – Boston – Berlin

© 2005 Birkhäuser Verlag, P.O. Box 133, CH-4010 Basel, Switzerland
Part of Springer Science+Business Media
Cover design: Birgit Blohmann, Zürich, Switzerland
Printed on acid-free paper produced from chlorine-free pulp. TCF ∞
Printed in Germany
ISBN-10: 3-7643-7134-X
ISBN-13: 978-3-7643-7134-0

9 8 7 6 5 4 3 2 1                                        www.birkhauser.ch

*Dedicated to my mother, to Daniel and Marina,*
*as well as to the memory of my father,*
*and of Abraham Stein, Haya Clara Stein (Bussika),*
*and Levy Welt.*

# Contents

# Introduction

As is well-known (and as the name strongly suggests), the Markov–Feller operators have appeared in the study of Feller processes, a type of Markov processes. These operators are used extensively in many areas such as, for example, in dynamical systems, in the study of iterated function systems with probabilities, and in the study of convolutions of measures.

In this work, we obtain "formulas" for supports of various types of invariant probabilities. Thus, we obtain such formulas for ergodic probabilities, for the unique invariant probability of a uniquely ergodic Markov–Feller operator, and for certain invariant probabilities that we call elementary and that have appeared in the work of Oxtoby and Ulam. In order to deal with the ergodic probabilities and their supports, we need a proper setting; so, we extend an ergodic decomposition which has emerged in the works of Krylov and Bogolioubov, Beboutoff, and Yosida. We call this decomposition the KBBY decomposition.

We think of this work as a natural outcome of [72]. In [72] we used topological lower limits in order to obtain a "formula" for the support of an attractive probability of a Markov–Feller operator; the "formula" was used to prove that, in a certain sense, the support of the attractive probability of a Markov–Feller operator defined by an iterated function system with place-dependent probabilities is independent of the place-dependent probabilities as long as these probabilities are strictly positive.

In June 1999 I discussed the results of [72] with Furstenberg, and he asked two questions concerning a Markov–Feller operator $T$ defined on a compact metric space $(X, d)$:

(1) If $T$ is uniquely ergodic, and $\mu$ is the unique $T$-invariant probability, is there a "formula" for the support of $\mu$ similar to the "formula" obtained in [72] for the support of an attractive probability?

(2) Assume that $T$ is not necessarily uniquely ergodic, but there exists a (closed nonempty) subset $F$ of $X$ such that supp $\mu = F$ for every $T$-invariant probability $\mu$ (that is, assume that all the $T$-invariant probabilities have the same support $F$). Is there a "formula" for $F$? This question was prompted by the following puzzling situation: one can construct a compact space $X$, and a continuous map $w : X \rightarrow X$ such that $w$ is minimal, and such that, if $T$ is the Markov–Feller operator induced by $w$, then $T$ is not uniquely ergodic; note that since $w$ is minimal, $X$ is the support of every $T$-invariant probability.

After having answered the above two questions (the answer to (1) is offered in Corollary 3.1.2, while the answer to (2) is in Theorem 3.1.3), we started to study the case in which the Markov–Feller operator is defined on a locally compact separable metric space; in this case, Example 1.1.14 "strongly suggests" an ergodic decomposition that has to be used in answering questions (1) and (2) in the locally compact case; later, while the final form of this work was being written, the ergodic decomposition that we obtained turned out to be the "push" needed to extend the KBBY decomposition; we call the decomposition that stems from Example 1.1.14 (example which was the starting point for obtaining our decomposition) a weak KBBY decomposition (see Section 2.1).

Although we have avoided the explicit use of topological limits in the statements and the proofs of our results, the "formulas" for supports were obtained by thinking in terms of these limits, and from time to time (for example, in Section 2.2 in the second paragraph after Corollary 2.2.3) we have pointed out our line of reasoning. Topological limits have been used before in studies related to Markov–Feller operators (although not as extensively and systematically as we once thought): when dealing with convolution of probabilities, these limits appear in the monograph by Heyer [28] in the case of convolutions on groups (see Theorem 2.1.4, pp. 91–92 of [28] and the comments on the history of the theorem at the end of Chapter 2 of [28]), and in the monograph by Högnäs and Mukherjea [29] in the case of convolutions on semigroups (note that Theorem 2.1.4 of [28], as well as the results of [29] can easily be stated in terms of convolution operators defined by probabilities, which are important examples of Markov–Feller operators); the limits were used in Theorem 2.7.1, pp. 37–39 of Barnsley [3], and in recent works by Lasota and Myjak ([37], [38], [39], [40], and [41]).

The work has four chapters. Except for the first chapter, the results of the remaining three chapters are new. The examples discussed in this work are used for illustrative purposes only.

As we just mentioned, Chapter 1 has an introductory character. Our goal in this chapter is to establish the notation, the terminology, the setting, and the results that will be used throughout the work. In order to improve the exposition, we define (in Section 1.1) the notion of a Markov–Feller pair: if $T$ is a Markov–Feller operator on measures, and if $S$ is the corresponding operator on functions, then the ordered pair $(S, T)$ is called a Markov–Feller pair. Since in this work we do not deal with Feller processes, we do not discuss the relationship between Markov–Feller operators and Feller processes. The interested reader can find a description of this relationship (actually, the more general relationship between transition probabilities and discrete-time Markov processes) in various monographs such as, for example, in Revuz [57], or the recent monograph of Hernández-Lerma and Lasserre [27].

We start Chapter 2 by extending the KBBY decomposition to any Markov–Feller operator (in Section 2.1). The extension is obtained by using the Lasota–Yorke lemma (Theorem 1.2.4) and the elementary measures that we mentioned earlier. Also in Section 2.1 we extend Theorem 1 of Oxtoby and Ulam [54] to our setting. In Section 2.2 we study the supports of the elementary measures, and we obtain "formulas" for the supports of the ergodic measures. Our investigation of the supports of the invariant measures is facilitated by extending the notion of orbit used in dynamical systems to Markov–Feller operators. In Section 2.3 we apply some of the results obtained in the first two sections to certain Markov–Feller operators called topologically connected by Skorokhod [64] (we also use the term "minimal" borrowed from dynamical systems). The results obtained in this section complement some results of Skorokhod [64] and extend a well-known theorem in ergodic theory (see Theorem 6.17 of Walters [67]).

In Chapter 3 we deal with uniquely ergodic Markov–Feller operators (that is, operators that have exactly one invariant probability). The "formulas" for the support of an ergodic measure obtained in Section 2.2 already yield a "formula" for the support of the unique invariant probability of a uniquely ergodic Markov–Feller operator; in Section 3.1 we obtain several additional "formulas" for the support of the invariant probability of a uniquely ergodic Markov–Feller operator, and we prove that if a Markov–Feller operator satisfies the conditions of the second question of Furstenberg, then all these "formulas" answer Furstenberg's second question. A topic of interest is to find simple criteria for unique ergodicity (see, for example, Hernández-Lerma and Lasserre [26]). In Section 3.2 we obtain such a criterion (another criterion is obtained in Section 4.1): we define a certain kind of generic points that we call dominant generic points, and we show that a Markov–Feller operator is uniquely ergodic if and only if the operator has at least one dominant generic point. In Section 3.3 we use the ideas of Section 3.2 to study ergodic measures. As a result of this study, we obtain alternative proofs for certain results of the classical KBBY decomposition.

As pointed out at various times in Chapter 2 and Chapter 3, the "formulas" for the support of the invariant probability of a uniquely ergodic Markov–Feller operator may yield a nonempty set, but the operator may fail to be uniquely ergodic. Thus, a natural goal is to find a large enough class of Markov–Feller operators such that an operator in that class is uniquely ergodic if and only if the "formulas" yield a nonempty set. In Chapter 4 we consider such a class, namely, the $C_0(X)$-equicontinuous Markov–Feller operators; in the compact case these operators emerged in the work of M. Rosenblatt [59] and [60]. In Section 4.1 we show that, indeed, a $C_0(X)$-equicontinuous Markov–Feller operator that has invariant probabilities is uniquely ergodic if and only if the "formulas" yield nonempty sets. Section 4.2 contains results needed to prove two ergodic theorems concerning $C_0(X)$-equicontinuous Markov–Feller operators; the two theorems and some easy consequences are discussed in Section 4.3.

The work is addressed to researchers interested in Markov–Feller operators, Feller processes, and related topics. However, it has been our goal to make the work accessible to newcomers to the above-mentioned areas. Thus, Chapter 1 is intended not only to establish the notation, the terminology, the setting, and the results that will be used throughout the work, but also to serve as a tutorial on Markov–Feller operators. Consequently, the work is self-contained in the sense that all the known results on Markov–Feller operators that are used throughout the volume are discussed in Chapter 1. The reader is assumed to be familiar with general topology, measure theory, and basic notions and results of functional analysis.

The topics of general topology and measure theory that we use here are covered by Cohn's book [8]; the notions and the results of general functional analysis that we need can be found in any textbook or monograph on the topic; finally, the elements of Banach lattice theory that will be encountered in this work can be found in any monograph that deals with Banach lattices, like for example, the books by Aliprantis and Burkinshaw [2], Schaefer [63], and Zaanen [71] (however, the reader unfamiliar with Banach lattices may simply ignore the places where we deal with ideas from the theory of Banach lattices without loosing relevant information about the results of this volume; we use elements of Banach lattice theory from time to time throughout the volume because the operators involved in the work are positive operators defined on rather standard types of Banach lattices, and a familiarity with Banach lattices and positive operators helps in understanding and simplifying certain ideas discussed in this volume). We believe that the work is also suitable for use in an advanced graduate course on Markov–Feller operators.

Finally, a word about the conventions used in the work: the labeling of formulas, theorems, propositions, corollaries, and lemmas is (we hope) self-explanatory; the end of a proof is denoted by the usual $\Box$, while the end of an example, remark, or observation is denoted by ∎.

# Acknowledgements

The pleasure of writing this work has been significantly enhanced by the support of a number of people: Onésimo Hernández-Lerma sent me his works, and has continuously encouraged me while writing the volume; Jean B. Lasserre, Arunava Mukherjea, and Murray Rosenblatt sent me their papers; Hillel Furstenberg has significantly influenced the results (as pointed out in the Introduction); Yitzhak Katznelson, Michael Lin, and Benji Weiss made valuable comments as we were discussing some of the results; Jon Aaronson brought to my attention that topological limits are treated in Kuratowski's monograph [34]; Rick Durrett and Harry Kesten offered me the opportunity to present some of the results of this work in the Probability Seminar at Cornell University; Şafak Alpay invited me to visit the Department of Mathematics at the Middle East Technical University in Ankara (where part of this book was written) and made every possible effort to make my stay a pleasant experience; Ersan Akyıldız, Aydın Aytuna, and the people in the Math Department in Ankara have created an environment that helped me a lot in my work; Andreas Tiefenbach has guided me as I was learning LATEX(this volume is the first document that I have ever typed myself)−actually, Andreas is to be credited with LATEXtraining a significant segment of the Turkish mathematical community through personal contacts and/or seminars; Eberhard Gerlach, Andrei Iacob, Patrick Ion, Jane Kister, and Marina Zaharopol went over the Introduction, and their suggestions have improved the exposition substantially; Norman Richert showed me how to access the TEXprograms as soon as I started to work at Mathematical Reviews; Marius Iosifescu showed a kind interest in this work, and suggested that I submit it for publication to Birkhäuser where Thomas Hempfling and his coworkers handled the publication process in an expert and efficient way; the anonymous referees made the effort and spent their time to evaluate the work, and one of the referees made recommendations that have significantly improved the book (Example 2.2.4, for instance, is due to him).

Naturally, I would like to express my gratitude to all of them.

# Chapter 1

# Preliminaries on Markov–Feller Operators

Our goal in this chapter is to discuss the definitions, the notations, and the known results in topology, functional analysis, and especially Markov–Feller operators that will be used throughout the volume.

In Section 1.1 we define the Markov–Feller pairs which, we believe, improve the exposition throughout the work. The discussion in this section is geared toward an essential result of M. Rosenblatt (Theorem 1.1.5) which states that every Markov–Feller pair is generated by a transition probability. We also discuss two recent results of Lasota and Myjak (Proposition 1.1.7 and Corollary 1.1.8) which will be used often. We conclude the section with several examples of Markov–Feller operators. The examples have a didactic purpose only; that is, the examples will just be used to illustrate the results of the next three chapters.

In Section 1.2 we review the various types of invariant probabilities of Markov–Feller pairs, a lemma of Lasota and Yorke that is instrumental in many places in this work, several results on almost everywhere convergence related to invariant probabilities, and a decomposition of Krylov, Bogolioubov, Beboutoff, and Yosida (the KBBY-decomposition).

Finally, in Section 1.3 we discuss briefly topological limits, Banach limits, a proof of the separability of $C_0(X)$ and several useful facts that stem from the proof in the case in which $X$ is a locally compact separable metric space ($C_0(X)$ stands for the Banach space of all real-valued continuous functions that vanish at infinity (see Section 1.1)), positive operators in vector lattices, with emphasis on the case when the vector lattices are Banach lattices, and equicontinuity.

## 1.1   Markov–Feller Pairs and Transition Probabilities

Given a nonempty set **S**, and a real-valued function $u$ defined on **S**, we say that $u$ is positive if $u(x) \geq 0$ for every $x \in \mathbf{S}$. In order to indicate that a function $u$ is positive, we use the notation $u \geq 0$. Observe that the function constant zero is a positive function in our terminology (as will be seen in the subsection *Vector Lattices, Banach Lattices, and Positive Operators* of Section 1.3, the terminology used here is in agreement with standard vector lattice terminology). Note that the set $\mathbb{R}$ of all real numbers can be thought of as a set of real-valued functions defined on **S** in the case in which **S** is a singleton (has only one element). In agreement with our terminology, a positive real number is a number $a \in \mathbb{R}$ such that $a = 0$ or $a > 0$. However, since the words "positive number" are sometimes used with the meaning "strictly positive number," we will refrain from using the words "$a$ is a positive number," and will use instead "$a$ is a nonnegative number" whenever $a \geq 0$; a similar wording convention applies when $a$ is a rational number, or an integer (as usual, the set of all natural numbers is the set of all strictly positive integers, and is denoted by $\mathbb{N}$).

Let $(X, d)$ be a locally compact separable metric space. (Throughout this volume $X$ stands for a locally compact separable metric space unless explicitly stated otherwise.) We will use the following notations:

$\mathcal{M}(X) =$ the Banach space of all real-valued signed Borel measures on $X$ (the norm on $\mathcal{M}(X)$ is the usual one, namely, the total variation norm);

$C_b(X) =$ the Banach space of all real-valued continuous bounded functions defined on $X$ (the norm on $C_b(X)$ is the usual sup (uniform) norm: $\| f \| = \sup_{t \in X} |f(t)|$ for every $f \in C_b(X)$);

$C_0(X) =$ the Banach space of all real-valued continuous functions that vanish at infinity (the norm on $C_0(X)$ is the usual sup (uniform) norm inherited from $C_b(X)$; that is, we think of $C_0(X)$ as a Banach subspace of $C_b(X)$);

$B_b(X) =$ the space of all real-valued Borel measurable bounded functions on $X$ (actually, $B_b(X)$ is a Banach space if we endow $B_b(X)$ with the sup (uniform) norm defined in the same way as the norm of $C_b(X)$; however, we do not need the Banach space structure of $B_b(X)$ in this book);

$\langle f, \mu \rangle = \int f(x) \, d\mu(x)$ for every $f \in B_b(X)$ and $\mu \in \mathcal{M}(X)$.

If $E$ is any of the spaces $C_0(X)$, $C_b(X)$, or $B_b(X)$, then the elements of $E$ are real-valued functions defined on $X$; if $u \in E$ is a positive function, then we will often call it a *positive element of $E$*, and, of course, in order to indicate that $u$ is a positive function, we will use the notation $u \geq 0$ (thus, for example, if $u \in C_b(X)$ and $u \geq 0$, we say that $u$ is a positive element of $C_b(X)$). Similar observations apply to the elements of $\mathcal{M}(X)$ since these elements are real-valued functions defined on the Borel $\sigma$-algebra of $X$; thus, if $\mu \in \mathcal{M}(X)$, we say that $\mu$ is a *positive element of* $\mathcal{M}(X)$ if $\mu(A) \geq 0$ for every Borel subset $A$ of $X$; we indicate that $\mu$ is a positive element of $\mathcal{M}(X)$ by using the notation $\mu \geq 0$. Observe that according to the above definition the zero measure (that is, the measure $\zeta$ defined by $\zeta(A) = 0$ for every Borel subset $A$ of $X$) is a positive element of $\mathcal{M}(X)$.

Now let $\mathbf{E}$ be any of the spaces $C_0(X)$, $C_b(X)$, $B_b(X)$, or $\mathcal{M}(X)$. A linear operator $Q : \mathbf{E} \to \mathbf{E}$ is called a *positive operator* if $Qw \geq 0$ for every positive element $w$ of $\mathbf{E}$. It can be shown that such a positive operator is bounded (continuous). The linear operator $Q : \mathbf{E} \to \mathbf{E}$ is called a *contraction* if $T$ is a bounded (continuous) operator, and $\|Q\| \leq 1$.

A positive contraction $T : \mathcal{M}(X) \to \mathcal{M}(X)$ is called a *Markov operator* if $\|T\mu\| = \|\mu\|$ for every $\mu \in \mathcal{M}(X)$, $\mu \geq 0$. There is a rather extensive literature on the various aspects of the theory of positive contractions and Markov operators (and the literature continues to grow at a fast pace). The interested reader may consult, for example, the books by Foguel [18], Hernández-Lerma and Lasserre [27], Krengel [32], Lasota and Mackey [36], Meyn and Tweedie [49], Nummelin [51], Orey [52], Revuz [57], the memoir by Szarek [66], and the paper by Diaconis and Freedman [14]. Another type of Markov operator will be defined in the subsection *Almost Everywhere Convergence Results* of Section 1.2. The two types of Markov operators are strongly connected (as shown in the above-mentioned subsection of Section 1.2), and the literature that we just mentioned deals with both types of Markov operators; actually, as we will see in the subsection *Vector Lattices, Banach Lattices, and Positive Operators* of Section 1.3, both types of Markov operators are particular cases of a more general notion of Markov operator.

Let $S : C_b(X) \to C_b(X)$ be a linear operator, and let $T : \mathcal{M}(X) \to \mathcal{M}(X)$ be a Markov operator. The pair $(S, T)$ is called a *Markov–Feller pair* if

$$\langle Sf, \mu \rangle = \langle f, T\mu \rangle \tag{1.1.1}$$

for every $f \in C_b(X)$ and $\mu \in \mathcal{M}(X)$.

A Markov operator $T : \mathcal{M}(X) \to \mathcal{M}(X)$ is called a *Markov–Feller operator* (or a *Feller operator*) if there exists a linear operator $S : C_b(X) \to C_b(X)$ such that $(S, T)$ is a Markov–Feller pair.

For every $x \in X$ we denote by $\delta_x$ the Dirac measure concentrated at $x$ (that is, $\delta_x$ is a probability in $\mathcal{M}(X)$ such that $\delta_x(\{x\}) = 1$).

Using Dirac measures it is easy to see that if $(S, T)$ is a Markov–Feller pair, then $S$ is a positive contraction of $C_b(X)$. Indeed,

$$Sf(x) = \langle Sf, \delta_x \rangle = \langle f, T\delta_x \rangle \geq 0$$

whenever $f \in C_b(X)$, $f \geq 0$, and $x \in X$. Consequently, $S$ is a positive operator. Since $S$ is positive, it is also a bounded operator. Let $1_X$ be the real-valued function defined on $X$ by $1_X(x) = 1$ for every $x \in X$ (in general, if $A$ is a subset of $X$, then we denote by $1_A$ the real-valued function on $X$ defined by $1_A(x) = 1$ whenever $x \in A$, and $1_A(x) = 0$ whenever $x \in X \backslash A$). Then $S1_X(x) = \langle 1_X, T\delta_x \rangle = 1$ for every $x \in X$; that is, $S1_X = 1_X$. Taking into consideration that $1_X$ is the largest element in the unit ball of $C_b(X)$ (that is, $1_X \geq f$ for every $f \in C_b(X)$, $\|f\| \leq 1$), and using the fact that $S$ is a positive operator, we obtain that $S$ is a contraction.

If $(X, d)$ is a compact space, then $C_b(X) = C_0(X)$, and $\mathcal{M}(X)$ is the dual of $C_b(X)$ $(= C_0(X))$; thus, if $X$ is compact, and $(S, T)$ is a Markov–Feller pair defined on $X$, then (1.1.1) implies that $T$ is the adjoint of $S$. However, if $X$ is not compact, it is no longer true that $T$ is the adjoint of $S$ (since $\mathcal{M}(X)$ fails to be the dual of $C_b(X)$). If $Sf \in C_0(X)$ whenever $f \in C_0(X)$ (that is, if $C_0(X)$ is an invariant subspace of $S$), then the restriction of $S$ to $C_0(X)$ can be thought of as a positive contraction of $C_0(X)$; in this case, $T$ can be thought of as the adjoint of the restriction of $S$ to $C_0(X)$. Unfortunately, in many (but not all) cases of interest $C_0(X)$ is not an invariant subspace of $S$.

Given a Markov–Feller pair $(S, T)$, even though it is not true, in general, that one of the operators is the adjoint of the other, the operators $S$ and $T$ have many properties that we would expect if one of the operators would have been the adjoint of the other (the reason that this happens is because $S$ and $T$ satisfy the equalities (1.1.1), of course). The next lemma deals with such a property.

**Lemma 1.1.1.** *If $(S, T)$ and $(S, T')$ are two Markov–Feller pairs defined on a locally compact separable metric space $(X, d)$, then $T = T'$.*

*Proof.* Assume that $T \neq T'$. Then $T\mu \neq T'\mu$ for some $\mu \in \mathcal{M}(X)$. Since $\mathcal{M}(X)$ is the dual of $C_0(X)$, there exists $f \in C_0(X)$ such that $\langle f, T\mu \rangle \neq \langle f, T'\mu \rangle$. We have obtained a contradiction since $\langle f, T\mu \rangle = \langle Sf, \mu \rangle$ and $\langle f, T'\mu \rangle = \langle Sf, \mu \rangle$.       □

Let $(X, d)$ be a locally compact separable metric space. We shall denote by $\mathcal{B}(X)$ (or, simply $\mathcal{B}$ if $X$ is understood) the $\sigma$-algebra of all Borel subsets of $X$.

A map $P : X \times \mathcal{B}(X) \to \mathbb{R}$ is called a *transition probability* if the following two conditions are satisfied:

(i) For every $x \in X$ the map $\mu_x : \mathcal{B}(X) \to \mathbb{R}$ defined by $\mu_x(A) = P(x, A)$ for every $A \in \mathcal{B}(X)$ is a probability measure.

(ii) For every $A \in \mathcal{B}(X)$ the function $g_A : X \to \mathbb{R}$ defined by $g_A(x) = P(x, A)$ for every $x \in X$ is Borel measurable.

Now assume that $P$ is a transition probability defined on $(X, d)$. For every $\mu \in \mathcal{M}(X)$ let $\bar{\mu} : \mathcal{B}(X) \to \mathbb{R}$ be defined by $\bar{\mu}(A) = \int P(x, A) \, d\mu(x)$ for every $A \in \mathcal{B}(X)$. Note that since the function $g_A = P(\cdot, A)$ is Borel measurable and $0 \leq P(x, A) \leq 1$ for every $x \in X$ and $A \in \mathcal{B}(X)$, it follows that $P(\cdot, A)$ is $\mu$-integrable for every $A \in \mathcal{B}(X)$; therefore, $\bar{\mu}(A)$ is well-defined. Using the monotone convergence theorem we obtain that if $\mu$ is a positive measure, then $\bar{\mu}$ is also a positive measure. Clearly, $\bar{\mu}(X) = \mu(X)$ whenever $\mu$ is a positive measure. Since every signed measure $\mu$ in $\mathcal{M}(X)$ is the difference of two positive elements of $\mathcal{M}(X)$, it follows that $\bar{\mu} \in \mathcal{M}(X)$ whenever $\mu \in \mathcal{M}(X)$.

From the above discussion it is easy to see that the map $T : \mathcal{M}(X) \to \mathcal{M}(X)$ defined by

$$T\mu(A) = \int P(x, A) \, d\mu(x) \tag{1.1.2}$$

for every $\mu \in \mathcal{M}(X)$ and $A \in \mathcal{B}(X)$ (that is, $T\mu = \bar{\mu}$ for every $\mu \in \mathcal{M}(X)$) is well defined. Clearly, $T$ is a linear operator, and, from our above comments, we obtain that $T$ is actually a Markov operator. We say that $T$ is the *Markov operator generated (or defined) by* $P$.

Let $T$ be a Markov operator generated by a transition probability $P$ (defined on $X$). One may ask at this point: is $T$ necessarily a Markov–Feller operator? In other words: is it true that there exists a positive contraction $S$ of $C_b(X)$ such that $(S, T)$ is a Markov–Feller pair?

In order to better understand the above question, let $T$ be a Markov operator generated by a transition probability $P$, and note that (1.1.2) implies that $T\delta_x(A) = P(x, A)$ for every $x \in X$ and $A \in \mathcal{B}(X)$. Thus, $T\delta_x = \mu_x$, where $\mu_x$ is the probability measure that appears in condition (i) of the definition of a transition probability.

Now assume that there exists $S$ such that $(S, T)$ is a Markov–Feller pair. Then

$$Sf(x) = \langle Sf, \delta_x \rangle = \langle f, T\delta_x \rangle = \int f(y) \, d\mu_x(y)$$

or

$$Sf(x) = \int f(y) \, P(x, dy) \tag{1.1.3}$$

for every $f \in C_b(X)$ and $x \in X$, where $P(x, dy)$ stands for $d\mu_x(y)$.

The above remarks show that if $(S, T)$ is a Markov–Feller pair, and if $T$ is generated by a transition probability $P$, then $S$ is defined by (1.1.3). If $T$ is a Markov operator defined by a transition probability $P$, it may well happen that $T$ fails to be a Markov–Feller operator, simply, because for some $f \in C_b(X)$ the function $Sf$ as defined by (1.1.3) may fail to be continuous. The next example illustrates such a situation:

*Example* 1.1.2. Let $X = [0, 2]$, assume that $X$ is endowed with the standard metric $d$ defined by $d(x, y) = |x - y|$ for every $x, y \in [0, 2]$, and let $\lambda$ be the Lebesgue measure on $[0, 2]$. We define $P : X \times \mathcal{B}(X) \to \mathbb{R}$ as follows:

$$P(x, A) = \begin{cases} \lambda(A \cap [0, 1]) & \text{if } x \in [0, 1] \\ \delta_x(A) = 1_A(x) & \text{if } x \in (1, 2] \end{cases}$$

for every $A \in \mathcal{B}(X)$.

Clearly, $P(x, \cdot)$ is a probability measure for every $x \in [0, 2]$. Since

$$P(x, A) = \lambda(A \cap [0, 1]) 1_{[0,1]}(x) + 1_{A \setminus [0,1]}(x)$$

for every $x \in [0, 2]$ and $A \in \mathcal{B}([0, 2])$, it follows that $P(\cdot, A)$ is a measurable function for every $A \in \mathcal{B}([0, 2])$. Thus, $P$ is a transition probability.

Let $T$ be the Markov operator generated by $P$. We will prove that $T$ cannot be a Markov–Feller operator. To this end, assume that $T$ is a Markov–Feller operator, let $S : C_b([0,2]) \to C_b([0,2])$ be such that $(S,T)$ is a Markov–Feller pair, and let $f : [0,2] \to \mathbb{R}$ be defined by

$$f(x) = \begin{cases} 1 - x & \text{if} \quad x \in [0,1] \\ 0 & \text{if} \quad x \in (1,2]. \end{cases}$$

Clearly, $f \in C_b([0,2])$. In view of our discussion of (1.1.3) we should be able to obtain $Sf$ by using (1.1.3). However, if we use (1.1.3), we obtain that

$$Sf(x) = \int f(y)P(x,\mathrm{d}y) = \int_0^1 (1-y)\,\mathrm{d}y = \frac{1}{2}$$

whenever $x \in [0,1]$, and

$$Sf(x) = \int f(y)P(x,\mathrm{d}y) = \int f(y)\,\mathrm{d}\delta_x(y) = f(x) = 0$$

whenever $x \in (1,2]$. We have obtained a contradiction since (1.1.3) does not yield a continuous function in this case. ■

So far, we have seen that given a transition probability $P$, we can always use $P$ and (1.1.2) to obtain a Markov operator $T$, and that, in general, $T$ is not a Markov–Feller operator. A natural question in this context is: given a Markov–Feller pair $(S,T)$, can we find a transition probability $P$ such that $T$ is defined by (1.1.2) (and $S$ is given by (1.1.3))? That is, is every Markov–Feller operator generated by a transition probability? Surprisingly, the answer is yes, and in the case in which the Markov–Feller pair is defined on a compact space, the proof appears on p. 118 of Rosenblatt's book [60]. Our goal now is to discuss Rosenblatt's proof in the locally compact case (in our setting). To this end, we need some preparation.

A topological space $Y$ is called $\sigma$-*compact* if there exists a sequence $(K_n)_{n \in \mathbb{N}}$ of compact subsets of $Y$ such that $Y = \bigcup\limits_{n=1}^{\infty} K_n$. Throughout the next proposition (and the entire work), given a locally compact separable metric space $(X,d)$, we will use the following standard notations:

$$B(x,r) = \{y \in X \mid d(x,y) < r\} = \text{the open ball in } X \text{ centered at } x \text{ of}$$
$$\text{radius } r, \text{ where } x \in X \text{ and } r \in \mathbb{R}, \ r > 0;$$
$$\bar{A} \text{ or } cl(A) = \text{the closure in } X \text{ of the subset } A \text{ of } X.$$

**Proposition 1.1.3.** *Every locally compact separable metric space $(X,d)$ is $\sigma$-compact.*

*Proof.* For every $x \in X$ set

$$\alpha_x = \sup\{r \in \mathbb{R} \mid r > 0 \text{ and } \overline{B(x,r)} \text{ is compact in } X\}.$$

Clearly, $\alpha_x > 0$ for every $x \in X$ since $X$ is locally compact.

If $\alpha_{x_0} = +\infty$ for some $x_0 \in X$, then $X$ is $\sigma$-compact since, in this case, $\overline{B(x_0, n)}$ is compact for every $n \in \mathbb{N}$, and $X = \bigcup\limits_{n=1}^{\infty} \overline{B(x_0, n)}$. Thus, we may assume that $\alpha_x < +\infty$ for every $x \in X$.

Since $X$ is separable, there exists a finite or countable subset $D$ of $X$ such that $D$ is dense in $X$. Clearly, the proof of the proposition will be completed if we show that $X = \bigcup\limits_{z \in D} \overline{B\left(z, \dfrac{\alpha_z}{2}\right)}$.

To this end, assume that $X \neq \bigcup\limits_{z \in D} \overline{B\left(z, \dfrac{\alpha_z}{2}\right)}$. Then, there exists $x \in X$ such that $x \notin \bigcup\limits_{z \in D} \overline{B\left(z, \dfrac{\alpha_z}{2}\right)}$. Since $D$ is dense in $X$ it follows that there exists a sequence $(z_k)_{k \in \mathbb{N}}$ of elements of $D$ such that $(z_k)_{k \in \mathbb{N}}$ converges to $x$ in the metric topology of $X$.

Since $x \notin \overline{B\left(z_k, \dfrac{\alpha_{z_k}}{2}\right)}$ for every $k \in \mathbb{N}$, it follows that $(\alpha_{z_k})_{k \in \mathbb{N}}$ converges to zero, and $d(x, z_k) \geq \dfrac{\alpha_{z_k}}{2}$ for every $k \in \mathbb{N}$.

Since $(z_k)_{k \in \mathbb{N}}$ converges to $x$, and since $(\alpha_{z_k})_{k \in \mathbb{N}}$ converges to zero, there exists $l \in \mathbb{N}$ large enough such that $d(x, z_l) < \dfrac{\alpha_x}{3}$ and $2\alpha_{z_l} < \dfrac{\alpha_x}{3}$.

If $y \in X$ is such that $d(y, z_l) \leq 2\alpha_{z_l}$, then

$$d(x, y) \leq d(x, z_l) + d(z_l, y) < \dfrac{\alpha_x}{3} + 2\alpha_{z_l} \leq \dfrac{2\alpha_x}{3};$$

consequently,

$$\{y \in X \,|\, d(y, z_l) \leq 2\alpha_{z_l}\} \subseteq B\left(x, \dfrac{2\alpha_x}{3}\right). \tag{1.1.4}$$

Clearly, the inclusion (1.1.4) implies that $\overline{B(z_l, 2\alpha_{z_l})} \subseteq \overline{B\left(x, \dfrac{2\alpha_x}{3}\right)}$.

We have obtained a contradiction since, as a result of the way in which $\alpha_{z_l}$ was picked, $\overline{B(z_l, 2\alpha_{z_l})}$ is a closed noncompact set included in the compact set $\overline{B\left(x, \dfrac{2\alpha_x}{3}\right)}$. $\qquad\square$

Let $P$ be a transition probability defined on a locally compact separable metric space $(X, d)$, and let $T$ be the Markov operator generated by $P$. We saw (in Example 1.1.2 and the discussion preceding the example) that it may happen that there exists $f \in C_b(X)$ such that the real-valued function $Sf$ defined on $X$ by (1.1.3) fails to be continuous, and that the existence of such an $f$ implies that $T$ fails to be a Markov–Feller operator. Thus, a natural question is: assume that $Sf$ is a continuous function whenever $f \in C_b(X)$ and $Sf$ is defined by (1.1.3);

hence (1.1.3) defines an operator $S : C_b(X) \to C_b(X)$. Is $(S, T)$ a Markov–Feller pair? The answer (which is obviously of interest in itself) is needed in order to prove that any Markov–Feller pair is generated by a transition probability, and is discussed in the next proposition.

**Proposition 1.1.4.** *Let $P$ be a transition probability defined on $(X, d)$, and let $T$ be the Markov operator generated by $P$. Assume that for every $f \in C_b(X)$ the function $\tilde{f} : X \to \mathbb{R}$ defined by $\tilde{f}(x) = \int f(y) \, P(x, dy)$ for every $x \in X$, is continuous. Then the map $S : C_b(X) \to C_b(X)$, $Sf = \tilde{f}$ for every $f \in C_b(X)$ is a well defined linear operator, and $(S, T)$ is a Markov–Feller pair.*

*Proof.* If $f \in C_b(X)$, then, clearly, $\tilde{f}$ is a bounded function (since $P(x, \cdot)$ is a probability on $(X, \mathcal{B}(X))$ for every $x \in X$); hence $\tilde{f} \in C_b(X)$. Thus, the map $S$ is well-defined (in the sense that $Sf \in C_b(X)$ for every $f \in C_b(X)$). Clearly, $S$ is linear.

We now prove that $(S, T)$ is a Markov–Feller pair.

To this end, we first note that if $g \in B_b(X)$, then $g$ is integrable with respect to any probability measure on $(X, \mathcal{B}(X))$; in particular, $g$ is integrable with respect to any of the probabilities $P(x, \cdot)$, $x \in X$; thus, it makes sense to define $\tilde{g} : X \to \mathbb{R}$, $\tilde{g}(x) = \int g(y) \, P(x, dy)$ for every $x \in X$. We now note that $\tilde{g}$ is (Borel) measurable for every $g \in B_b(X)$. (Indeed, if $g = 1_A$ for some $A \in \mathcal{B}(X)$, then $\tilde{g}(x) = P(x, A)$ for every $x \in X$, so, $\tilde{g}$ is clearly measurable in this case. Consequently, $\tilde{g}$ is measurable whenever $g$ is a simple function. In general, if $g \in B_b(X)$, then there exists a sequence $(g_n)_{n \in \mathbb{N}}$ of simple Borel measurable functions such that $(g_n)_{n \in \mathbb{N}}$ converges uniformly to $g$; therefore, $\tilde{g}_n$ is measurable for every $n \in \mathbb{N}$, and it is easy to see that the sequence $(\tilde{g}_n)_{n \in \mathbb{N}}$ converges uniformly to $\tilde{g}$; accordingly, $\tilde{g}$ is measurable.) Since $\tilde{g}$ is obviously a bounded function whenever $g \in B_b(X)$, it follows that $\tilde{g} \in B_b(X)$ for every $g \in B_b(X)$. Thus, we can extend $S$ to $B_b(X)$ (the extension is denoted by $S$, as well since it is easy to see from the context which one of the operators is under consideration) as follows: $S : B_b(X) \to B_b(X)$ is defined by $Sg = \tilde{g}$ for every $g \in B_b(X)$.

In order to prove that $(S, T)$ is a Markov–Feller pair we only have to prove that the operators $T$ and $S$ satisfy (1.1.1) for every $\mu \in \mathcal{M}(X)$ and $f \in C_b(X)$. To this end, we will actually prove a bit more: we will show that $T$ and the extended operator $S$ satisfy (1.1.1) for every $f \in B_b(X)$ and $\mu \in \mathcal{M}(X)$.

Now let $\mu \in \mathcal{M}(X)$ be fixed. We have to prove that given $S$, $T$, and $\mu$, it follows that (1.1.1) is true for every $f \in B_b(X)$. If $f = 1_A$ for some $A \in \mathcal{B}(X)$, then

$$\langle Sf, \mu \rangle = \iint 1_A(y) P(x, dy) \, d\mu(x) = \int P(x, A) \, d\mu(x) = \langle f, T\mu \rangle.$$

Therefore, (1.1.1) holds true for every simple function $f$. Since for every $f \in B_b(X)$ there exists a sequence of simple functions which converges uniformly to $f$, it is easy to see that (1.1.1) is valid for every $f \in B_b(X)$. □

If $f : X \to \mathbb{R}$ is a continuous function, supp $f$ denotes the *topological support of $f$*; that is, supp $f = \overline{\{x \in X | f(x) \neq 0\}}$. We will denote by $C_c(X)$ the vector subspace of $C_0(X)$ of all functions with compact supports.

Proposition 1.1.3 and Proposition 1.1.4 allow us to discuss M. Rosenblatt's theorem and his proof that we mentioned earlier.

**Theorem 1.1.5 (M. Rosenblatt).** *Any Markov–Feller pair defined on a locally compact separable metric space is generated by a transition probability; that is, given a Markov–Feller pair $(S, T)$ defined on a locally compact separable metric space $(X, d)$, there exists a transition probability $P$ on $(X, d)$ such that $T$ is defined by $(1.1.2)$ and $S$ is defined by $(1.1.3)$.*

*Proof.* Let $x \in X$, and consider the map $\mu_x : C_0(X) \to \mathbb{R}$ defined by $\mu_x(f) = Sf(x)$ for every $f \in C_0(X)$.

Clearly, $\mu_x$ is a positive linear functional on $C_0(X)$; therefore, $\mu_x$ is also continuous. Since (using the Riesz representation theorem) we can identify the dual of $C_0(X)$ with $\mathcal{M}(X)$, we may and do think of $\mu_x$ as a positive measure in $\mathcal{M}(X)$.

Let $P : X \times \mathcal{B}(X) \to \mathbb{R}$ be defined by $P(x, A) = \mu_x(A)$ for every $x \in X$ and $A \in \mathcal{B}(X)$.

Obviously, in order to complete the proof of the theorem we have to show that:

(i) $P$ is a transition probability.

(ii) The Markov–Feller pair $(S, T)$ is generated by $P$.

(i) In order to prove that $P$ is a transition probability we have to show that:

(i)-(a) $P(x, \cdot)$ is a probability measure for every $x \in X$.

(i)-(b) $P(\cdot, A)$ is a measurable function for every $A \in \mathcal{B}(X)$.

(i)-(a) Let $x \in X$. Since $P(x, \cdot) = \mu_x$ is a positive measure, in order to prove that $P(x, \cdot)$ is a probability, we only have to show that $\mu_x(X) = 1$.

Clearly,

$$\mu_x(X) = \|\mu_x\| = \sup_{\substack{f \in C_0(X) \\ \|f\| \leq 1}} |\langle f, \mu_x \rangle| = \sup_{\substack{f \in C_0(X) \\ \|f\| \leq 1}} |Sf(x)| \leq 1 \qquad (1.1.5)$$

since $S$ is a positive contraction of $C_b(X)$ (because $(S, T)$ is a Markov–Feller pair).

Since the metric space $(X, d)$ is $\sigma$-compact (by Proposition 1.1.3), there exists a sequence $(K_n)_{n \in \mathbb{N}}$ of compact subsets of $X$ such that $K_n \subseteq K_{n+1}$ for every $n \in \mathbb{N}$ and such that $X = \bigcup_{m=1}^{\infty} K_m$.

By Proposition 7.1.8, p. 199 of Cohn's book [8], there exists $f_n \in C_0(X)$ (actually, we can choose $f_n$ to be a continuos function with compact support) such that $1_{K_n} \leq f_n \leq 1_X$ for every $n \in \mathbb{N}$.

Since $(f_n)_{n\in\mathbb{N}}$ converges pointwise to $1_X$, and since $1_X$ is integrable with respect to the probability measure $T\delta_x$, we can apply the Lebesgue dominated convergence theorem in order to conclude that

$$\lim_{n\to\infty} \langle f_n, T\delta_x \rangle = \int 1_X(y)\, \mathrm{d}(T\delta_x)(y) = 1.$$

Since

$$\langle f_n, \mu_x \rangle = Sf_n(x) = \langle Sf_n, \delta_x \rangle = \langle f_n, T\delta_x \rangle$$

for every $n \in \mathbb{N}$, we obtain that

$$\lim_{n\to\infty} \langle f_n, \mu_x \rangle = 1. \tag{1.1.6}$$

Clearly, (1.1.5) and (1.1.6) imply that $\mu_x(X) = 1$.

(i)-(b) Set

$$\mathcal{A} = \{A \in \mathcal{B}(X) | P(\cdot, A) \text{ is a measurable function}\}.$$

In order to prove that $P(\cdot, A)$ is a measurable function whenever $A \in \mathcal{B}(X)$, it is obviously enough to prove that $\mathcal{B}(X) \subseteq \mathcal{A}$.

We first prove that $G \in \mathcal{A}$ whenever $G$ is an open subset of $X$.

To this end, let $G$ be an open subset of $X$. In order to show that $P(\cdot, G)$ is measurable, we will prove that the set $A_\alpha = \{x \in X | P(x, G) > \alpha\}$ is open in $X$ whenever $\alpha \in \mathbb{R}$.

Thus, let $\alpha \in \mathbb{R}$. Since the empty set is open we may assume that $A_\alpha \neq \emptyset$.

Let $x \in A_\alpha$. Using the proof of the Riesz representation theorem (see, for example, formula (4) in the proof of Theorem 7.2.8, pp. 209–210 of Cohn's book [8]), we obtain that

$$P(x, G) = \mu_x(G) = \sup_{\substack{f \in C_c(X) \\ supp\, f \subseteq G \\ 0 \le f \le 1}} \langle f, \mu_x \rangle = \sup_{\substack{f \in C_c(X) \\ supp\, f \subseteq G \\ 0 \le f \le 1}} Sf(x).$$

Accordingly, there exists $f \in C_0(X)$, $supp\, f \subseteq G$, $0 \le f \le 1$ such that $Sf(x) > \alpha$. The set $U_x = \{z \in X | Sf(z) > \alpha\}$ is open since $Sf$ is a continuous function. Moreover, it is easy to see that $x \in U_x$ and that $U_x \subseteq A_\alpha$. Consequently, $A_\alpha$ is an open set whenever $\alpha \in \mathbb{R}$; hence, $G \in \mathcal{A}$ whenever $G$ is an open set.

It is easy to see that:

($\star$) For every $n \in \mathbb{N}$ and $n$ disjoint subsets $A_1, A_2, \ldots, A_n$ of $X$ such that $A_i \in \mathcal{A}$ for every $i = 1, 2, \ldots, n$, it follows that $\bigcup_{j=1}^{n} A_j \in \mathcal{A}$ (that is, $\mathcal{A}$ is closed under the formation of finite disjoint unions).

($\star\star$) If $A, B \in \mathcal{A}$, $A \supseteq B$, then $A \setminus B \in \mathcal{A}$ (that is, $\mathcal{A}$ is closed under the formation of proper differences).

($\star\star\star$) If $(A_n)_{n\in\mathbb{N}}$ is a sequence of elements of $\mathcal{A}$ such that $(A_n)_{n\in\mathbb{N}}$ is monotone (that is, $(A_n)_{n\in\mathbb{N}}$ is increasing ($A_n \subseteq A_{n+1}$ for every $n \in \mathbb{N}$), or decreasing ($A_n \supseteq A_{n+1}$ for every $n \in \mathbb{N}$)), then $\lim\limits_{n\to\infty} A_n \in \mathcal{A}$ where $\lim\limits_{n\to\infty} A_n = \bigcup\limits_{n=1}^{\infty} A_n$ if $(A_n)_{n\in\mathbb{N}}$ is increasing, and $\lim\limits_{n\to\infty} A_n = \bigcap\limits_{n=1}^{\infty} A_n$ if $(A_n)_{n\in\mathbb{N}}$ is decreasing (that is, $\mathcal{A}$ is a monotone class).

Let $\mathcal{R}$ be the class of all finite disjoint unions of proper differences of compact subsets of $X$. Clearly, $\mathcal{R} \subseteq \mathcal{A}$ because the compact subsets of $X$ belong to $\mathcal{A}$ (to see that the compact subsets of $X$ belong to $\mathcal{A}$ note that all the open subsets of $X$ belong to $\mathcal{A}$; therefore, ($\star\star$) implies that all the closed subsets of $X$ belong to $\mathcal{A}$).

By Theorem F, p. 223 of the book by Halmos [23] the class $\mathcal{R}$ is a ring (that is, $A \cup B \in \mathcal{R}$ and $A \setminus B \in \mathcal{R}$ whenever $A \in \mathcal{R}$ and $B \in \mathcal{R}$).

Let $M(\mathcal{R})$ be the monotone class generated by $\mathcal{R}$ (that is, the smallest monotone class that contains $\mathcal{R}$; such a class exists since the class of all subsets of $X$ is obviously monotone, and an intersection of monotone classes is again a monotone class). By Theorem B, p. 27 of the book by Halmos [23], $M(\mathcal{R})$ is closed under the formation of countable unions, and under the formation of differences (that is, $\bigcup\limits_{n=1}^{\infty} A_n \in M(\mathcal{R})$ whenever $(A_n)_{n\in\mathbb{N}}$ is a sequence of subsets of $X$ such that $A_n \in M(\mathcal{R})$ for every $n \in \mathbb{N}$, and $A \setminus B \in M(\mathcal{R})$ whenever $A \in M(\mathcal{R})$ and $B \in M(\mathcal{R})$). Since $X$ is $\sigma$-compact by Proposition 1.1.3, it follows that $X \in M(\mathcal{R})$; hence, $M(\mathcal{R})$ is a $\sigma$-algebra.

Let $(K_n)_{n\in\mathbb{N}}$ be an increasing sequence of compact subsets of $X$ such that $X = \bigcup\limits_{n=1}^{\infty} K_n$. If $F$ is a closed subset of $X$, then $F = \bigcup\limits_{n=1}^{\infty} (F \cap K_n)$, and $(F \cap K_n)_{n\in\mathbb{N}}$ is an increasing sequence of compact subsets of $X$; therefore, $F \in M(\mathcal{R})$. Since every closed subset of $X$ belongs to $M(\mathcal{R})$ and $M(\mathcal{R})$ is a $\sigma$-algebra, it follows that $\mathcal{B}(X) \subseteq M(\mathcal{R})$. Since $\mathcal{A}$ is a monotone class that includes $\mathcal{R}$, we obtain that $M(\mathcal{R}) \subseteq \mathcal{A}$; consequently, $\mathcal{B}(X) \subseteq \mathcal{A}$.

(ii) We first note that in order to prove that the Markov–Feller pair $(S, T)$ is generated by $P$, it is enough to prove that $S$ and $P$ satisfy (1.1.3) for every $f \in C_b(X)$ and $x \in X$. Indeed, assume that (1.1.3) is satisfied, and let $T'$ be the Markov operator generated by $P$; by Proposition 1.1.4, the pair $(S, T')$ is a Markov–Feller pair; since both $(S, T)$ and $(S, T')$ are Markov–Feller pairs, Lemma 1.1.1 implies that $T = T'$.

Since $P(x, \cdot) = \mu_x$ for every $x \in X$, it follows that

$$Sf(x) = \int f(y)\, d\mu_x(y) = \int f(y) P(x,\, dy)$$

for every $f \in C_0(X)$ and $x \in X$; thus, (1.1.3) is satisfied whenever $f \in C_0(X)$ (and $x \in X$).

Now let $f \in C_b(X)$. Since $X$ is $\sigma$-compact, there exists an increasing sequence $(K_n)_{n\in\mathbb{N}}$ of compact subsets of $X$ such that $X = \bigcup_{n=1}^{\infty} K_n$. By Proposition 7.1.8, p. 199 of Cohn's book [8], for every $n \in \mathbb{N}$ there exists a function $g_n$ with compact support such that $1_{K_n} \leq g_n \leq 1_X$. Set $f_n = fg_n$ for every $n \in \mathbb{N}$. Then $f_n \in C_0(X)$ for every $n \in \mathbb{N}$, and $(f_n)_{n\in\mathbb{N}}$ converges pointwise to $f$ (that is, $(f_n(x))_{n\in\mathbb{N}}$ converges to $f(x)$ for every $x \in X$). By the Lebesgue dominated convergence theorem we have

$$Sf(x) = \langle Sf, \delta_x \rangle = \langle f, T\delta_x \rangle = \lim_{n\to\infty}\langle f_n, T\delta_x \rangle = \lim_{n\to\infty} Sf_n(x)$$
$$= \lim_{n\to\infty} \int f_n(y)P(x,\,\mathrm{d}y) = \int f(y)P(x,\,\mathrm{d}y)$$

for every $x \in X$. Thus, (1.1.3) holds true whenever $f \in C_b(X)$ and $x \in X$.  □

Let $(S, T)$ be a Markov–Feller pair defined on a locally compact separable metric space $(X, d)$, and let $S^n$ and $T^n$, $n \in \mathbb{N}$ be the iterates of $S$ and $T$, respectively. It is easy to see that $(S^n, T^n)$ is a Markov–Feller pair for every $n \in \mathbb{N}$; by Theorem 1.1.5 $(S^n, T^n)$ is generated by a transition probability which we denote by $P_n$, $n \in \mathbb{N}$.

**Proposition 1.1.6.** *If $(S, T)$ is a Markov–Feller pair defined on $(X, d)$, and if $(S^n, T^n)$ is generated by the transition probability $P_n$, $n \in \mathbb{N}$, then*

$$P_{m+n}(x, A) = \int P_n(y, A)P_m(x,\,\mathrm{d}y) \tag{1.1.7}$$

*for every $m \in \mathbb{N}$, $n \in \mathbb{N}$, $x \in X$, and $A \in \mathcal{B}(X)$.*

*Proof.* We have already pointed out in the proof of Proposition 1.1.4 that $S$ can be extended to $B_b(X)$, and that the extended operator $S$ has the property that $\langle Sf, \mu \rangle = \langle f, T\mu \rangle$ for every $f \in B_b(X)$ and $\mu \in \mathcal{M}(X)$.

Thus,
$$\langle S^{m+n}f, \mu \rangle = \langle S^n f, T^m \mu \rangle \tag{1.1.8}$$

for every $m \in \mathbb{N}$, $n \in \mathbb{N}$, $f \in B_b(X)$, and $\mu \in \mathcal{M}(X)$.

Now let $x \in X$ and $A \in \mathcal{B}(X)$. In view of the comments preceding Example 1.1.2, we obtain that $T^l\delta_x(A) = P_l(x, A)$, and $S^l 1_A(x) = \langle S^l 1_A, \delta_x \rangle = P_l(x, A)$ for every $l \in \mathbb{N}$.

Using (1.1.8) we obtain that

$$P_{m+n}(x, A) = \langle S^{m+n}1_A, \delta_x \rangle = \langle S^n 1_A, T^m \delta_x \rangle$$
$$= \iint 1_A(z)P_n(y,\,\mathrm{d}z)P_m(x,\,\mathrm{d}y) = \int P_n(y, A)P_m(x,\,\mathrm{d}y)$$

for every $m \in \mathbb{N}$ and $n \in \mathbb{N}$.

Thus, (1.1.7) holds true whenever $m \in \mathbb{N}$, $n \in \mathbb{N}$, $x \in X$, and $A \in \mathcal{B}(X)$.  □

The next technical result and its corollary appear in Lasota and Myjak (see Proposition 3.1 and Corollary 3.1 of [38]), and will be used often throughout the volume. As usual, supp $\mu$ stands for the support of $\mu$ whenever $\mu \in \mathcal{M}(X)$; note that supp $\mu$ is well defined in our setting since we assume that $X$ is a locally compact separable metric space (see p. 226 of Cohn's book [8]).

**Proposition 1.1.7 (Lasota and Myjak, [38]).** *Let $(S, T)$ be a Markov–Feller pair defined on $(X, d)$. If $\mu \in \mathcal{M}(X)$, $\mu \geq 0$ and $\nu \in \mathcal{M}(X)$, $\nu \geq 0$ are such that supp $\mu \subseteq$ supp $\nu$, then supp $(T\mu) \subseteq$ supp $(T\nu)$.*

**Corollary 1.1.8 (Lasota and Myjak, [38]).** *If $(S, T)$ and $(X, d)$ are as in Proposition 1.1.7, then supp $(T\mu) =$ supp $(T\nu)$ whenever $\mu \in \mathcal{M}(X)$ and $\nu \in \mathcal{M}(X)$ are such that $\mu \geq 0$, $\nu \geq 0$, and supp $\mu =$ supp $\nu$.*

We will conclude this section with several examples of Markov–Feller operators.

As usual, let $(X, d)$ be a locally compact separable metric space, and let $w : X \to X$ be a continuous function. Using $w$ we can construct a Markov–Feller pair $(S, T)$ as follows: $T : \mathcal{M}(X) \to \mathcal{M}(X)$ is defined by $T\mu(A) = \mu(w^{-1}(A))$ for every $\mu \in \mathcal{M}(X)$ and $A \in \mathcal{B}(X)$, while $S : C_b(X) \to C_b(X)$ is defined by $Sf = f \circ w$ for every $f \in C_b(X)$. It is easy to see that $T$ and $S$ are well-defined (in the sense that $T\mu \in \mathcal{M}(X)$ for every $\mu \in \mathcal{M}(X)$ and $Sf \in C_b(X)$ for every $f \in C_b(X)$), and that $T$ is a Markov operator. Let $P : X \times \mathcal{B}(X) \to \mathbb{R}$ be defined by $P(x, A) = \delta_{w(x)}(A)$ for every $x \in X$ and $A \in \mathcal{B}(X)$. Using the equality $\delta_{w(x)}(A) = 1_{w^{-1}(A)}(x)$, $x \in X$, $A \in \mathcal{B}(X)$, we obtain that $P$ is a transition probability. It is easy to see that (1.1.2) is satisfied by $T$ and $P$ for every $\mu \in \mathcal{M}(X)$ and $A \in \mathcal{B}(X)$, and that (1.1.3) is satisfied by $S$ and $P$ for every $f \in C_b(X)$ and $x \in X$. By Proposition 1.1.4 we infer that $(S, T)$ is a Markov–Feller pair. We call $(S, T)$ the *Markov–Feller pair induced by $w$* (or, we say that $(S, T)$ is *induced by a continuous function*). Markov–Feller pairs induced by continuous functions are used in topological dynamics (see, for example, Chapter 3 of Furstenberg's book [20]).

*Example* 1.1.9. Let $X = \mathbb{N}$, and let $d$ be the usual metric on $\mathbb{N}$ ($d(i, j) = |i - j|$ for every $i \in \mathbb{N}$ and $j \in \mathbb{N}$). It is well-known and easy to prove directly that there exists a standard isometry from $\mathcal{M}(\mathbb{N})$ onto $l^1$ = the Banach space of all real-valued sequences $(\xi_n)_{n \in \mathbb{N}}$ such that $\sum_{n=1}^{\infty} |\xi_n| < \infty$ (the norm on $l^1$ is defined by $\|(\xi_n)_{n \in \mathbb{N}}\| = \sum_{n=1}^{\infty} |\xi_n|$ for every $(\xi_n)_{n \in \mathbb{N}} \in l^1$) that maps a Dirac measure $\delta_n$ to the element $1_{\{n\}}$ of $l^1$, $n \in \mathbb{N}$. Thus, we can think of the elements of $\mathcal{M}(\mathbb{N})$ as elements of $l^1$, and vice versa. Similarly, we can think of $C_b(\mathbb{N})$ as $l^\infty$ = the Banach space of all real-valued bounded sequences endowed with the usual norm of $l^\infty$ ($\|(\xi_n)_{n \in \mathbb{N}}\| = \sup_{n \in \mathbb{N}} |\xi_n|$ for every $(\xi_n)_{n \in \mathbb{N}} \in l^\infty$), and of $C_0(\mathbb{N})$ as $c_0$ = the Banach space of all real-valued sequences that converge to zero (the norm on $c_0$ is the restriction of the norm on $l^\infty$ to $c_0$). Note that the dual of $l^1$ is $l^\infty$; however, the dual of $l^\infty$ cannot be thought of as $l^1$ because, roughly speaking, there exist elements of the dual of $l^\infty$ (like, for example, the Banach limits (see Section 1.3 for details on Banach limits)) which cannot be thought of as elements of $l^1$. Any Markov operator $T : l^1 \to l^1$ is actually a Markov–Feller operator; indeed, if $S : l^\infty \to l^\infty$ is the adjoint of $T$, then $(S, T)$ is a Markov–Feller pair. ∎

*Example 1.1.10.* Let $X = \mathbb{N}$ as in Example 1.1.9, and let $w : \mathbb{N} \to \mathbb{N}$ be defined by $w(n) = n + 1$ for every $n \in \mathbb{N}$. If $(S, T)$ is the Markov–Feller pair induced by $w$, then $S : l^\infty \to l^\infty$ is defined by $S((a_1, a_2, a_3, \ldots)) = (a_2, a_3, a_4, \ldots)$ for every $(a_n)_{n\in\mathbb{N}} \in l^\infty$, and $T : l^1 \to l^1$ is defined by $T((a_1, a_2, a_3, \ldots)) = (0, a_1, a_2, a_3, \ldots)$ for every $(a_n)_{n\in\mathbb{N}} \in l^1$. The operator $S$ is usually referred to as a shift on $\mathbb{N}$; both $S$ and $T$ appear in literature often. ∎

*Example 1.1.11 (Rotations of the Unit Circle).* Let $X = \mathbb{R}/\mathbb{Z}$ be the unit circle thought of as a compact metric space and as a group (the algebraic operation that defines a group structure on $\mathbb{R}/\mathbb{Z}$ is the addition modulo 1). Let $a \in \mathbb{R}/\mathbb{Z}$ and consider the rotation of the circle by $a$; that is, consider the map $w_a : \mathbb{R}/\mathbb{Z} \to \mathbb{R}/\mathbb{Z}$ defined by $w_a(x) = x + a$ mod 1 for every $x \in \mathbb{R}/\mathbb{Z}$. Clearly, $w_a$ is a continuous function. If $(S_a, T_a)$ is the Markov–Feller pair induced by $w_a$, then $S_a : C_b(X) \to C_b(X)$ is defined by $S_a f(x) = f(x \oplus a)$ for every $f \in C_b(X)$ and $x \in X$, where $x \oplus a = x + a$ mod 1, $x \in X$, and $T_a : \mathcal{M}(X) \to \mathcal{M}(X)$ is defined by $T_a\mu = \mu(A \ominus a)$ for every $\mu \in \mathcal{M}(X)$ and $A \in \mathcal{B}(X)$, where $A \ominus a = \{y \in X \mid y = x - a$ mod 1 for some $x \in A\}$ for every $A \in \mathcal{B}(X)$.

The rotations of the unit circle are discussed in almost every monograph on ergodic theory. Furstenberg has extended these rotations to tori in order to obtain a new proof of Weyl's equidistribution theorem (see [20], Chapter 3, Section 3). More general transformations of tori are studied in the papers by Dani and Muralidharan [13], and Furstenberg [19]. We have included the rotations of the unit circle here because we will use them to illustrate several results later in the work. ∎

Let $(X, d)$ be a locally compact separable metric space. If $Y$ is a closed subset of $X$ and if $d_Y$ is the restriction of $d$ to $Y$, then it is well-known and easy to prove that $(Y, d_Y)$ is also a locally compact separable metric space.

Let $w : X \to X$ be a continuous function. A subset $A$ of $X$ is called *invariant* (or *w-invariant* if there is any danger of confusion) whenever $w(A) \subseteq A$.

Now let $Y$ be a closed invariant subset of $X$. Since the range of the restriction $w_Y$ of $w$ to $Y$ is included in $Y$, we may think of $w_Y$ as a function from $Y$ to $Y$. Since $w_Y : Y \to Y$ is continuous, we may consider the Markov–Feller pair $(S_Y, T_Y)$ induced by $w_Y$. It is often the case (as in Example 1.1.11 and Example 1.1.12 below) that along with the Markov–Feller pair $(S, T)$ induced by $w$, one also studies Markov–Feller pairs induced by restrictions of $w$ to certain closed invariant subsets of $X$.

*Example 1.1.12.* Let $X = \mathbb{R}$, and let $d$ be the usual metric on $\mathbb{R}$ ($d(x, y) = |x - y|$ for every $x \in \mathbb{R}$ and $y \in \mathbb{R}$). Let $a \in \mathbb{R}$, $a > 0$ and let $w_a : \mathbb{R} \to \mathbb{R}$ be defined by $w_a(x) = ax(1 - x)$ for every $x \in \mathbb{R}$. The maps $w_a$, $a > 0$ are known as quadratic maps or logistic maps, and have been studied extensively (see, for example, the monographs by Falconer [17], Lasota and Mackey [36], and Robinson [58], and the papers by Henry [24], Lasota and Mackey [35], Li and Yorke [43] and [44], May [47], Misiurewicz [50], Pianigiani [56], and Ruelle [62]). For every $a \in \mathbb{R}$ one can define the Markov–Feller pair $(S_a, T_a)$ induced by $w_a$. Of special interest is the case when $a \in (0, 4]$ since in this case the interval $[0, 1]$ is $w_a$-invariant; hence, we can consider the function $w_{a[0,1]}$ and the Markov–Feller pair $(S_{a[0,1]}, T_{a[0,1]})$ induced by $w_{a[0,1]}$. ∎

As before, let $(X, d)$ be a locally compact separable metric space, and let $w : X \to X$ be a continuous function. A subset $Y$ of $X$ is called *minimal* (with respect to $w$) if $Y$ is a closed nonempty invariant subset of $X$ such that the following condition is satisfied: if $Z$ is a closed nonempty invariant subset of $Y$, then $Z = Y$.

*Example* 1.1.13. Let $\Lambda$ be a finite set. In order to avoid trivialities, assume that $\Lambda$ has at least two elements, and, to simplify the notation, assume that $\Lambda = \{0, 1, 2, \ldots, l-1\}$. On $\Lambda$ we consider the metric $d_\Lambda$ defined by $d_\Lambda(i, j) = |i - j|$ for every $i \in \Lambda$ and $j \in \Lambda$. Let $X = \Lambda^{\mathbb{N}} = $ the set of all sequences of elements of $\Lambda$. On $X$ we define the metric $d$ as follows

$$d((i_k)_{k \in \mathbb{N}}, (j_k)_{k \in \mathbb{N}}) = \sum_{k=1}^{\infty} 2^{-k} d_\Lambda(i_k, j_k)$$

for every $(i_k)_{k \in \mathbb{N}} \in X$ and $(j_k)_{k \in \mathbb{N}} \in X$. Clearly, the topology generated by $d_\Lambda$ on $\Lambda$ is the collection $\mathcal{P}(\Lambda)$ of all the subsets of $\Lambda$, and the topology generated by $d$ on $X$ is the product topology. Now let $w : X \to X$ be defined by $w((i_1, i_2, i_3, i_4, \ldots )) = (i_2, i_3, i_4, i_5, \ldots )$ for every $(i_k)_{k \in \mathbb{N}} \in X$, and let $(S, T)$ be the Markov–Feller pair induced by $w$. Given a closed $w$-invariant subset $Y$ of $X$, the pair $(Y, w_Y)$ is called a *symbolic flow*. If $Y$ is a minimal subset of $X$ (with respect to $w$), then $(Y, w_Y)$ is called a *minimal symbolic flow*. A symbolic flow $(Y, w_Y)$ induces a Markov–Feller pair $(S_Y, T_Y)$. We also can consider symbolic flows of bisequences; that is, we can define $X = \Lambda^{\mathbb{Z}} = $ the topological space of all $\Lambda$-valued functions defined on the set $\mathbb{Z}$ of all integers. In this case, the metric $d$ is defined by $d((i_k)_{k \in \mathbb{Z}}, (j_k)_{k \in \mathbb{Z}}) = \lim_{n \to \infty} \sum_{k=-n}^{n} 2^{-|k|} d_\Lambda(i_k, j_k)$ for every $(i_k)_{k \in \mathbb{Z}} \in X$ and $(j_k)_{k \in \mathbb{Z}} \in X$. The map $w : X \to X$ is defined by $(w((i_k)_{k \in \mathbb{Z}}))_l = i_{l+1}$ for every $(i_k)_{k \in \mathbb{Z}} \in X$ and $l \in \mathbb{Z}$ where $(w((i_k)_{k \in \mathbb{Z}}))_l$ stands for the $l$th coordinate of $w((i_k)_{k \in \mathbb{Z}})$.

There is a rather large literature on symbolic flows. Here, we have followed the approach of Furstenberg [20]. The papers by Boshernitzan [6] and [7] deal with topics in the study of symbolic flows which are related to certain results that we discuss in this work. As pointed out in [6] any minimal interval exchange tranformation of a finite number of intervals (for the definition of these transformations, see the monograph by Cornfeld, Fomin, and Sinai [9]) has an isomorphic representation as a minimal symbolic flow; accordingly, many results for minimal interval exchange transformations yield similar results for minimal symbolic flows; typical examples of such a situation are the results of Keynes and Newton [31] and of Keane [30]. Symbolic flows, and, in particular, the full symbolic flow $(X, w)$ are convenient tools for the study of the iterates of various continuous maps, like, for example, the quadratic maps described in Example 1.1.12 (see the monographs by de Melo and van Strien [48], and Robinson [58]). Other topics in the study of symbolic flows (including applications in combinatorics (graph theory) and computer science (the storage of data)) can be found in the monograph by Lind and Marcus [46], and the papers by Coven [10], Coven and Hedlund [11], Coven and Paul [12], Durand [16], Goetz [21], and Paul [55].                             ■

We conclude this section with an example of a Markov–Feller pair that is not induced by a continuous function. As one no doubt expects, most Markov–Feller pairs are not induced by continuous functions; some examples are the Markov–Feller pairs generated by a transition probability of a right (or left) random walk generated by a probability measure (see Section 3.1 of the monograph by Högnäs and Mukherjea [29]), or the Markov–Feller pairs generated by iterated function systems with probabilities (see Lasota and Myjak [37], [38], [39], [40], [41] and Zaharopol [72], [73]).

*Example* 1.1.14. Let $X = \mathbb{N}$ and let $d$ be as in Example 1.1.9. Let $T : l^1 \to l^1$ be defined as follows: if $\alpha \in l^1$, $\alpha = (x_n)_{n\in\mathbb{N}}$, then $T\alpha = (y_n)_{n\in\mathbb{N}}$ where

$$
y_n = \begin{cases}
x_1 + x_3 & \text{if } n = 1 \\
0 & \text{if } n = 2 \\
x_{2k+1} & \text{if } n = 2k - 1, \ k \geq 2 \\
x_{2k-2} & \text{if } n = 2k, \ k \geq 2
\end{cases}.
$$

Thus, $T((x_n)_{n\in\mathbb{N}}) = (x_1 + x_3, \ 0, \ x_5, \ x_2, \ x_7, \ x_4, \ x_9, \ x_6, \ \ldots)$ for every $(x_n)_{n\in\mathbb{N}} \in l^1$.

Let $S : l^\infty \to l^\infty$ be the adjoint of $T$. Clearly, $(S, T)$ is a Markov–Feller pair (see Example 1.1.9).

Note that $S$ acts as follows: if $\gamma \in l^\infty$, $\gamma = (u_n)_{n\in\mathbb{N}}$, then $S\gamma = (v_n)_{n\in\mathbb{N}}$ where

$$
v_n = \begin{cases}
u_1 & \text{if } n = 1 \\
u_{2k-1} & \text{if } n = 2k + 1, \ k \geq 1 \\
u_{2k+2} & \text{if } n = 2k, \ k \geq 1
\end{cases};
$$

that is, $S((u_n)_{n\in\mathbb{N}}) = (u_1, \ u_4, \ u_1, \ u_6, \ u_3, \ u_8, \ u_5, \ \ldots)$ for every $(u_n)_{n\in\mathbb{N}} \in l^\infty$.

■

## 1.2 Invariant Probabilities

In the previous section we have introduced the notion of Markov–Feller pairs in order to improve the exposition in the work, and we used these pairs in order to outline the known results on Markov–Feller operators that are needed throughout the volume. In this section we review the known types of invariant probabilities of a Markov–Feller operator, a lemma of Lasota and Yorke (Lemma 3.1 of [42]), the Hopf ergodic theorem (see, for example, Theorem 3.5, pp. 128-129 of Krengel's book [32]) along with several results of Chapter 4 of Revuz's monograph [57], and an ergodic decomposition that we call the KBBY decomposition since it has emerged from the works of Krylov and Bogolioubov [33], Beboutoff [5], and Yosida [68] and [69] (we will follow Section 4 of Chapter 13 of Yosida's book [70]).

**Invariant Probabilities of Markov–Feller Operators.** Let $(X, d)$ be a locally compact separable metric space.

We will denote by $Pr(X)$ the set of all probability measures in $\mathcal{M}(X)$; that is, $Pr(X) = \{\mu \in \mathcal{M}(X)|\ \mu \geq 0 \text{ and } \|\mu\| = 1\}$.

If $\mu \in \mathcal{M}(X)$, we will use the standard notations $\mu^+ = \mu \vee 0$ and $\mu^- = (-\mu) \vee 0$. It is well known that $\mu = \mu^+ - \mu^-$.

Let $(S, T)$ be a Markov–Feller pair defined on $(X, d)$.

A measure $\mu \in \mathcal{M}(X)$ is called an *invariant measure for $T$* (or a *$T$-invariant measure*, or an *invariant measure for $(S,\ T)$*) if $T\mu = \mu$. Since the zero measure is always invariant, we are interested in nonzero invariant measures.

Assume that $T$ has nonzero invariant measures, and let $\mu \in \mathcal{M}(X)$ be such a nonzero $T$-invariant measure. Since $T$ is a positive operator, it follows that $T(\mu^+) \geq (T\mu)^+ = \mu^+$ and $T(\mu^-) \geq (T\mu)^- = \mu^-$; since $T$ is a Markov operator we obtain that $T(\mu^+) = \mu^+$ and $T(\mu^-) = \mu^-$. Thus, $\mu \in \mathcal{M}(X)$ is $T$-invariant if and only if $\mu^+$ and $\mu^-$ are $T$-invariant.

If $\mu \in \mathcal{M}(X)$ is a positive nonzero $T$-invariant measure, then $\dfrac{\mu}{\|\mu\|}$ is the unique element of $Pr(X)$ which is at the same time $T$-invariant and a scalar multiple of $\mu$.

The above discussion shows that for most purposes, in order to understand the structure of the set of all $T$-invariant elements, it is enough to understand the structure of the set of all $T$-invariant probabilities. That is why the topic of this work is the study of invariant probabilities of Markov–Feller operators.

It is well-known (see, for example, p. 178 of Krengel's book [32]) that if $(X, d)$ is compact, then the Markov–Feller pair $(S, T)$ has invariant probabilities. By contrast, if $(X, d)$ is a noncompact locally compact separable metric space, the Markov–Feller pair $(S, T)$ may or may not have invariant probabilities (the Markov–Feller pair in Example 1.1.10 does not have invariant probabilities, while the Markov–Feller pair in Example 1.1.14 does have one, namely the Dirac measure concentrated at 1).

We say that $T$ (or $(S,T)$) is *uniquely ergodic* if $T$ has exactly one invariant probability measure. We call $T$ (or $(S,T)$) *strictly ergodic* if $T$ (or $(S,T)$) is uniquely ergodic, and the support of the unique $T$-invariant probability is the entire space $X$. The Markov–Feller pair in Example 1.1.14 is uniquely ergodic but not strictly ergodic. If the equivalence class $a \in \mathbb{R}/\mathbb{Z}$ in Example 1.1.11 contains irrational numbers, then the corresponding Markov–Feller pair $(S_a, T_a)$ is strictly ergodic because the only invariant probability for $(S_a, T_a)$ is the Haar measure on $\mathbb{R}/\mathbb{Z}$ (it is customary to think of $\mathbb{R}/\mathbb{Z}$ as the interval $[0,1)$ in $\mathbb{R}$; in this case, the Haar measure on $\mathbb{R}/\mathbb{Z}$ can be thought of as the Lebesgue measure on $[0,1)$).

Let $\mu^* \in Pr(X)$. We say that $\mu^*$ is an *attractive probability* for $T$ (or for $(S,T)$) if the sequence $((\langle f, T^n\mu\rangle)_{n\in\mathbb{N}\cup\{0\}}$ converges to $\langle f, \mu^*\rangle$ whenever $f \in C_b(X)$ and $\mu \in Pr(X)$. It is easy to see that if $T$ has an attractive probability $\mu^*$, then $\mu^*$ is the unique $T$-invariant probability, so $T$ is uniquely ergodic. The study of attractive probabilities is of particular interest in the case in which the Markov–Feller pair $(S,T)$ is induced by an iterated function system with probabilities (see, for example, Barnsley, Demko, Elton, and Geronimo [4], or Lasota and Yorke [42]).

Let $P$ be the transition probability that generates the Markov–Feller pair $(S,T)$ (the existence of $P$ was proved in Theorem 1.1.5). A subset $A$ of $X$, $A \in \mathcal{B}(X)$ is called a *P-invariant set* (or an *invariant set with respect to P* if $P(x, A) = 1$ whenever $x \in A$. A measure $\mu \in \mathcal{M}(X)$ is called an *ergodic measure* (a *T-ergodic measure*, an *(S,T)-ergodic measure*, or an *ergodic measure for T* (or *for (S,T)*) if $\mu$ is a $T$-invariant probability, and $\mu(A) = 0$ or $1$ whenever $A \in \mathcal{B}(X)$ is a $P$-invariant set.

The above definition of ergodic measures is from Hernández-Lerma and Lasserre [25]. If the Markov–Feller pair $(S,T)$ is induced by a continuous function, then the definition of [25] is the same as the usual one for such Markov–Feller pairs (see, for example, the monographs by Furstenberg [20], Krengel [32], or the paper by Oxtoby [53]).

The following two theorems are consequences of results of Hernández-Lerma and Lasserre [25].

**Theorem 1.2.1.** *If the Markov–Feller pair $(S,T)$ has invariant probabilities, then $(S,T)$ has ergodic measures.*

**Theorem 1.2.2.** *The Markov–Feller pair $(S,T)$ is uniquely ergodic if and only if $(S,T)$ has a unique ergodic measure. Thus, if $(S,T)$ is uniquely ergodic, then the unique $T$-invariant probability is also an ergodic measure.*

**The Lasota–Yorke Lemma.** We will now discuss a result of Lasota and Yorke (Lemma 3.1 of [42]), which will be used often in this work. As before, we assume given a Markov–Feller pair $(S,T)$ defined on a locally compact separable metric space $(X,d)$.

Given $\mu \in \mathcal{M}(X)$, let $\tilde{\mu} : C_b(X) \to \mathbb{R}$ be defined by $\tilde{\mu}(f) = \int f(x)\, d\mu(x)$ for every $f \in C_b(X)$. We call $\tilde{\mu}$ the *standard extension of $\mu$ (to $C_b(X)$)*. It is easy to see that $\tilde{\mu}$ is a bounded linear functional on $C_b(X)$. It is also easy to see that if $\tilde{\mu}(Sf) = \tilde{\mu}(f)$ for every $f \in C_0(X)$, then $T\mu = \mu$. (Indeed, $\langle f, T\mu \rangle = \langle Sf, \mu \rangle = \tilde{\mu}(Sf) = \tilde{\mu}(f) = \langle f, \mu \rangle$ for every $f \in C_0(X)$ implies that $T\mu = \mu$.) The Lasota–Yorke lemma is an extension of the above observation to arbitrary positive linear functionals on $C_b(X)$. More precisely, let $\phi : C_b(X) \to \mathbb{R}$ be a positive linear functional. Then $\phi$ is bounded (continuous), so by the Riesz representation theorem we may think of the restriction $\mu_\phi$ of $\phi$ to $C_0(X)$ as an element of $\mathcal{M}(X)$. The Lasota–Yorke lemma states that if $\phi(Sf) = \phi(f)$ for every $f \in C_0(X)$, and if $\phi$ satisfies an additional mild condition, then $\mu_\phi$ is $T$-invariant.

In order to discuss the Lasota–Yorke lemma in detail we need some preparation.

In the next lemma and throughout the volume we will denote by $d(x, A)$ the *distance from a point* $x \in X$ *to a subset* $A$ *of* $X$ (that is, $d(x, A) = \inf_{y \in A} d(x, y)$).

**Lemma 1.2.3.** *Let* $h \in C_b(X)$, $h \geq 0$, *and set* $\mathcal{H} = \{u \in C_0(X) |\ 0 \leq u \leq h\}$. *Then* $\sup \mathcal{H} = h$ *where the supremum is taken in* $C_b(X)$.

*Proof.* Let $h \in C_b(X)$, $h \geq 0$, and set $\mathcal{H} = \{u \in C_0(X) \mid 0 \leq u \leq h\}$.

Clearly, $h$ is an upper bound for $\mathcal{H}$. Assume that $h$ is not the least upper bound of $\mathcal{H}$. Then there exist $f \in C_b(X)$ and $x_0 \in X$ such that $f(x_0) < h(x_0)$ and such that $f$ is an upper bound for $\mathcal{H}$. By Proposition 7.1.8, p. 199 of Cohn [8], there exists $g \in C_c(X)$ such that $0 \leq g \leq 1_X$ and $g(x_0) = 1$. Since $hg \in \mathcal{H}$ (so, $hg \leq f$), it follows that $h(x_0) = (hg)(x_0) \leq f(x_0)$. We have obtained a contradiction which stems from the assumption that $h$ is not the least upper bound of $\mathcal{H}$. Accordingly, $h = \sup \mathcal{H}$. $\qquad\square$

We are now ready to prove the Lasota–Yorke lemma.

**Theorem 1.2.4 (Lasota–Yorke Lemma).** *Let* $\phi : C_b(X) \to \mathbb{R}$ *be a positive linear functional such that* $\phi(1_X) = 1$ *and* $\phi(Sf) = \phi(f)$ *for every* $f \in C_0(X)$. *Then the restriction* $\mu_\phi$ *of* $\phi$ *to* $C_0(X)$ *has the property that* $T\mu_\phi = \mu_\phi$ *provided that we think of* $\mu_\phi$ *as an element of* $\mathcal{M}(X)$.

*Proof.* Let $\tilde{\mu}_\phi$ be the standard extension of $\mu_\phi$ to $C_b(X)$.

We first note that $\mu_\phi(h) \leq \phi(h)$ for every $h \in C_b(X)$, $h \geq 0$. Indeed, if $h \in C_b(X)$, $h \geq 0$, and if $\mathcal{H} = \{u \in C_0(X) |\ 0 \leq u \leq h\}$, then Lemma 1.2.3 of this section and Proposition 7.4.4, pp. 229–230 of Cohn's monograph [8] imply that $\sup_{u \in \mathcal{H}} \tilde{\mu}_\phi(u) = \tilde{\mu}_\phi(h)$. Since $\phi$ is a positive functional, it follows that

$$\tilde{\mu}_\phi(h) = \sup_{u \in \mathcal{H}} \tilde{\mu}_\phi(u) = \sup_{u \in \mathcal{H}} \phi(u) \leq \phi(h).$$

Now, $\langle f, T\mu_\phi \rangle = \langle Sf, \mu_\phi \rangle = \tilde{\mu}_\phi(Sf) \leq \phi(Sf) = \phi(f) = \langle f, \mu_\phi \rangle$ for every $f \in C_0(X)$. Accordingly, $T\mu_\phi = \mu_\phi$. $\qquad\square$

Note that if $\phi$ and $\mu_\phi$ are as in Theorem 1.2.4, then $\mu_\phi \geq 0$ and $\|\mu_\phi\| \leq 1$ since $0 \leq \langle f, \mu_\phi \rangle = \phi(f) \leq \phi(1_X) = 1$ for every $f \in C_0(X)$, $0 \leq f \leq 1_X$.

**Almost Everywhere Convergence Results.** In this subsection we review briefly several topics on almost everywhere (a.e.) convergence. We start with the Hopf ergodic theorem; although we state the theorem in a less general form that suits our needs better, it still preserves the full flavor of Hopf's theorem (for a more general version, a proof, and additional details, see Theorem 3.5, pp. 128-129 and Section 3.3 of Krengel's book [32]).

Let $(Y, \Sigma, \mu)$ be a measure space, and let $\mathbf{M}(Y, \Sigma, \mu)$ be the vector space of all classes of equivalence of real-valued measurable functions on $Y$ (two functions $f_1$ and $f_2$ belong to the same class if and only if $f_1 = f_2$ $\mu$-almost everywhere ($\mu$-a.e.)). As usual, we say that a sequence $(\overline{f}_n)_{n\in\mathbb{N}}$ of elements of $\mathbf{M}(Y, \Sigma, \mu)$ converges $\mu$-a.e. if there exists $\overline{f} \in \mathbf{M}(Y, \Sigma, \mu)$ such that for every sequence $(g_n)_{n\in\mathbb{N}}$ of real-valued measurable functions on $Y$ where $g_n$ belongs to $\overline{f}_n$ for all $n \in \mathbb{N}$, and for every function $g$ in the class $\overline{f}$ the sequence $(g_n)_{n\in\mathbb{N}}$ converges $\mu$-a.e. to $g$. Clearly, $(\overline{f}_n)_{n\in\mathbb{N}}$ converges $\mu$-a.e. if and only if there exist $\overline{f} \in \mathbf{M}(Y, \Sigma, \mu)$, a sequence $(h_n)_{n\in\mathbb{N}}$ of functions where $h_n$ is in the class $\overline{f}_n$ for all $n \in \mathbb{N}$, and a function $h$ in the class $\overline{f}$ such that $(h_n)_{n\in\mathbb{N}}$ converges $\mu$-a.e. to $h$.

Let $p \in \mathbb{R} \cup \{\infty\}$, $1 \leq p \leq +\infty$ and let $L^p(Y, \Sigma, \mu)$ be the usual Banach space. As we all know, if $\overline{f} \in L^p(Y, \Sigma, \mu)$, then we say that $\overline{f}$ is a positive element of $L^p(Y, \Sigma, \mu)$ if there exists a real-valued measurable function $g$ in the class $\overline{f}$ such that $g \geq 0$ $\mu$-a.e. A linear operator $T : L^p(Y, \Sigma, \mu) \to L^p(Y, \Sigma, \mu)$ is called a *positive operator* if $T\overline{f} \geq 0$ whenever $\overline{f}$ is a positive element of $L^p(Y, \Sigma, \mu)$. The linear operator $T$ is called a *contraction* (of $L^p(Y, \Sigma, \mu)$) if $T$ is bounded (continuous), and $\|T\| \leq 1$.

Let $T : L^1(Y, \Sigma, \mu) \to L^1(Y, \Sigma, \mu)$ be a positive contraction. We say that $T$ is a *Markov operator* if $\|T\overline{f}\| = \|\overline{f}\|$ whenever $\overline{f} \in L^1(Y, \Sigma, \mu)$, $\overline{f} \geq 0$. Note that the Markov operators introduced here are different from the Markov operators defined at the beginning of Section 1.1; however, as we will point out in the subsection *Vector Lattices, Banach Lattices, and Positive Operators* of Section 1.3, one can define a more general type of Markov operators such that the Markov operators defined here and in Section 1.1 are particular cases of the Markov operators defined in Section 1.3.

Given a Banach space $E$, and a linear operator $T : E \to E$, set $A_n(T)u = \frac{1}{n} \sum_{k=0}^{n-1} T^k u$ for every $n \in \mathbb{N}$ and $u \in E$.

**Theorem 1.2.5 (The Hopf Ergodic Theorem).** *Let $(Y, \Sigma, \mu)$ be a probability space, and let $T : L^1(Y, \Sigma, \mu) \to L^1(Y, \Sigma, \mu)$ be a positive contraction such that $T\overline{1}_Y = \overline{1}_Y$. Then:*

(a) *The sequence $(A_n(T)\overline{f})_{n\in\mathbb{N}}$ converges $\mu$-a.e., and the $\mu$-a.e. limit is an element of $L^1(Y, \Sigma, \mu)$ whenever $\overline{f} \in L^1(Y, \Sigma, \mu)$.*

(b) *If $\overline{f} \in L^1(Y, \Sigma, \mu)$ and $\overline{g}$ is the $\mu$-a.e. limit of $(A_n(T)\overline{f})_{n\in\mathbb{N}}$, then $\int \overline{f} \, d\mu = \int \overline{g} \, d\mu$.*

The next theorem summarizes various results of Chapter 4 of Revuz's book [57].

**Theorem 1.2.6.** *Let $(S,T)$ be a Markov–Feller pair defined on a locally compact separable metric space $(X, d)$, and let $\mu \in \mathcal{M}(X)$ be a $T$-invariant probability. If $f \in C_0(X)$, then the sequence $\left( \frac{1}{n} \sum_{k=0}^{n-1} S^k f \right)_{n \in \mathbb{N}}$ converges $\mu$-a.e. to a $\mu$-integrable function $g$, and $\langle f, \mu \rangle = \int g \, d\mu$.*

*Proof.* It is well-known that the Radon–Nikodym theorem implies that the set of all the elements of $\mathcal{M}(X)$ that are absolutely continuous with respect to the measure $\mu$ is a Banach subspace of $\mathcal{M}(X)$ that is isometric to $L^1(X, \mathcal{B}(X), \mu)$. By Lemma 5.1 of Lin [45] (or by Proposition 1.1 of Chapter 4 of the monograph by Revuz [57]) the operator $T$ defines another operator $U : L^1(X, \mathcal{B}(X), \mu) \to L^1(X, \mathcal{B}(X), \mu)$ as follows: $U(\overline{f})$ is the Radon–Nikodym derivative of $T(\overline{f}\mu)$ with respect to $\mu$. It is easy to see that $U$ is a Markov operator on $L^1(X, \mathcal{B}(X), \mu)$.

Let $U' : L^\infty(X, \mathcal{B}(X), \mu) \to L^\infty(X, \mathcal{B}(X), \mu)$ be the dual of $U$. By Proposition 1.4, Chapter 4 of Revuz [57] we have $U'\overline{f} = \overline{Sf}$ for every $f \in B_b(X)$ where $\overline{g}$ denotes the class of Borel measurable functions equal $\mu$-a.e. to $g$. Note that by Rosenblatt's theorem (Theorem 1.1.5) the Markov–Feller pair $(S,T)$ is generated by a transition probability; therefore, (as pointed out in the proof of Proposition 1.1.4) the operator $S$ can be extended to $B_b(X)$, so $Sf$ is well-defined whenever $f \in B_b(X)$.

Since the equality (1.1.1) holds true for every $f \in B_b(X)$ and every element of $\mathcal{M}(X)$, we obtain that $\int U'\overline{1}_A \, d\mu = \langle S1_A, \mu \rangle = \langle 1_A, T\mu \rangle = \mu(A)$ for every $A \in \mathcal{B}(X)$; therefore, $U'$ can be extended to an operator $V : L^1(X, \mathcal{B}(X), \mu) \to L^1(X, \mathcal{B}(X), \mu)$.

It is easy to see that $V$ is a Markov operator, and that

$$\overline{1}_X = \overline{S1_X} = U'\overline{1}_X = V\overline{1}_X.$$

Thus, we can apply the Hopf ergodic theorem (Theorem 1.2.5) to $V$.

Now let $f \in C_0(X)$. By Theorem 1.2.5 the sequence $\left( \frac{1}{n} \sum_{k=0}^{n-1} V^k \overline{f} \right)_{n \in \mathbb{N}}$ converges $\mu$-a.e. to some $\overline{g}$, $\overline{g} \in L^1(X, \mathcal{B}(X), \mu)$, and $\int \overline{g} \, d\mu = \int f \, d\mu$.

Since $V^k \overline{f} = \overline{S^k f}$ for every $k \in \mathbb{N} \cup \{0\}$, it follows that $\left( \frac{1}{n} \sum_{k=0}^{n-1} S^k f \right)_{n \in \mathbb{N}}$ converges $\mu$-a.e. to $g$. $\qquad \square$

**Corollary 1.2.7.** *Let $(S, T)$ and $\mu$ be as in Theorem 1.2.6, and let $f \in C_0(X)$ be such that $f \geq 0$ and $\langle f, \mu \rangle > 0$. Set*

$$\Theta = \left\{ x \in X \;\middle|\; \left( \frac{1}{n} \sum_{k=0}^{n-1} S^k f(x) \right)_{n \in \mathbb{N}} \text{ converges and } \lim_{n \to \infty} \frac{1}{n} \sum_{k=0}^{n-1} S^k f(x) > 0 \right\}.$$

*Then:*

(a) $\mu(\Theta) > 0$.

(b) $\Theta \cap supp\ \mu \neq \emptyset$.

*Proof.* (a) By Theorem 1.2.5 there exists a $\mu$-integrable function $g$ such that the sequence $\left( \frac{1}{n} \sum_{k=0}^{n-1} S^k f \right)_{n \in \mathbb{N}}$ converges $\mu$-a.e. to $g$ and $\int g\, d\mu = \int f\, d\mu > 0$. Accordingly, $\mu(\Theta) > 0$ since $\{g > 0\} = \Theta$ $\mu$-a.e.

(b) $\Theta \cap supp\ \mu \neq \emptyset$ because $\mu(\Theta) > 0$ and $\mu(X \setminus (supp\ \mu)) = 0$.  □

**The KBBY Decomposition.** As mentioned at the beginning of this section, the decomposition that we will discuss now is the result of the works of Krylov and Bogolioubov [33], Beboutoff [5], and Yosida [68] and [69] (see also Hernández-Lerma and Lasserre [25]). The idea of such a decomposition has appeared in Krylov and Bogolioubov [33]; the case of a Markov–Feller pair $(S, T)$ defined on a compact metric space $(X, d)$ was dealt with by Beboutoff [5] and Yosida [68]. Finally, Yosida was able in [69] to extend the decomposition to the case in which $(X, d)$ is a metric space whose closed bounded sets are compact, $P : X \times \mathcal{B}(X) \rightarrow \mathbb{R}$ is a transition probability, and $S : B_b(X) \rightarrow B_b(X)$ is defined by (1.1.3) and maps $C_c(X)$ into $C_c(X)$. (Note that the proof of Proposition 1.1.4 shows that $S$ is well-defined.) The features of the decomposition will "come to life" in Chapter 2 where we will show that these features are preserved in our setting.

Let $(S, T)$ be a Markov–Feller pair defined on a locally compact separable metric space $(X, d)$.

Set

$$\mathcal{D}(S, T) = \left\{ x \in X \ \middle| \ \begin{array}{l} \text{the sequence } (A_n(S)f(x))_{n \in \mathbb{N}} \text{ converges to zero} \\ \text{whenever } f \in C_0(X) \end{array} \right\},$$

$$\Gamma_0(S, T) = X \setminus \mathcal{D}(S, T),$$

and

$$\Gamma_c(S, T) = \left\{ x \in \Gamma_0(S, T) \ \middle| \ \begin{array}{l} \text{the sequence } (A_n(S)f(x))_{n \in \mathbb{N}} \text{ converges} \\ \text{for every } f \in C_0(X) \end{array} \right\}.$$

If $x \in \Gamma_c(S, T)$, then it makes sense to define $\varepsilon_x : C_0(X) \rightarrow \mathbb{R}$ by $\varepsilon_x(f) = \lim_{n \to \infty} A_n(S)f(x)$ for every $f \in C_0(X)$. Clearly, $\varepsilon_x$ is a positive linear functional on $C_0(X)$, so $\varepsilon_x \in \mathcal{M}(X)$; it is also obvious that $0 < \|\varepsilon_x\| \leq 1$.

Now set $\Gamma_{cp}(S, T) = \{ x \in \Gamma_c(S, T) \mid \|\varepsilon_x\| = 1 \}$.

The set $\mathcal{D}(S, T)$ is called the *dissipative part of $X$ (generated by $(S, T)$)*; the Markov–Feller pair is called *dissipative* if $\mathcal{D}(S, T) = X$. In general, it will be clear from the context which Markov–Feller pair is under consideration, so we will use the notations $\mathcal{D}$, $\Gamma_0$, $\Gamma_c$, and $\Gamma_{cp}$ rather than $\mathcal{D}(S, T)$, $\Gamma_0(S, T)$, $\Gamma_c(S, T)$, and $\Gamma_{cp}(S, T)$, respectively. As the reader familiar with Yosida's work might expect, the set $\Gamma_{cp}$ will be decomposed further in Chapter 2.

If $(S,T)$ is a Markov–Feller pair defined on a compact metric space $(X,d)$, then $\mathcal{D}$ is empty since $A_n(S)1_X = 1_X$ for every $n \in \mathbb{N}$. By contrast, if $X$ is not compact, then $(S,T)$ may well be dissipative (see Example 1.1.10).

Of course, one can find a Markov–Feller pair $(S,T)$ (in the noncompact case) such that both $\mathcal{D}$ and $\Gamma_0$ are nonempty. Such a situation appears in Example 1.1.14. If $X = \mathbb{N}$ and $(S,T)$ is the Markov–Feller pair of Example 1.1.14, then $\mathcal{D} = \{2k \mid k \in \mathbb{N}\}$ while $\Gamma_0 = \{2k - 1 \mid k \in \mathbb{N}\}$.

The sets $\Gamma_0$ and $\Gamma_c$ can be distinct even in the compact case. To illustrate this point, let $(S,T)$ be the Markov–Feller pair of Example 1.1.13 where $\Lambda = \{0,1\}$, and $X = \Lambda^{\mathbb{N}}$. Now let $f : X \to \mathbb{R}$ be defined by $f((i_k)_{k\in\mathbb{N}}) = i_1$ for every $(i_k)_{k\in\mathbb{N}} \in X$. It is easy to find an element $\alpha = (i_k)_{k\in\mathbb{N}}$ of $X$ such that the sequence $\left(\frac{1}{n}\sum_{k=0}^{n-1} S^k f(\alpha)\right)_{n\in\mathbb{N}}$ does not converge. Since $f$ is continuous and $X$ is compact it follows that $f \in C_0(X)$, so $\alpha \notin \Gamma_c$; however, $\alpha \in \Gamma_0$ because the compactness of $X$ implies that $\Gamma_0 = X$. Note that in this example $\Gamma_c \neq \emptyset$ since the sequence $(0, 0, 0, \ldots)$ belongs to $\Gamma_c$.

It is easy to see that in the compact case $\Gamma_c = \Gamma_{cp}$. However, (in the non-compact case) the sets $\Gamma_c$ and $\Gamma_{cp}$ can be distinct as illustrated in the following slight modification of Example 1.1.14.

*Example* 1.2.8. Let $X = \mathbb{N}$ as in Example 1.1.14, and let $T : l^1 \to l^1$ be defined as follows: if $\alpha \in l^1$, $\alpha = (x_n)_{n\in\mathbb{N}}$, then $T\alpha = (y_n)_{n\in\mathbb{N}}$ where

$$y_n = \begin{cases} x_1 + \frac{1}{2}x_2 + x_3 & \text{if } n = 1 \\ 0 & \text{if } n = 2 \\ x_{2k+1} & \text{if } n = 2k - 1, \ k \geq 2 \\ \frac{1}{2}x_2 & \text{if } n = 4 \\ x_{2k-2} & \text{if } n = 2k, \ k \geq 3 \end{cases}.$$

Thus,

$$T((x_n)_{n\in\mathbb{N}}) = (x_1 + \frac{1}{2}x_2 + x_3, \ 0, \ x_5, \ \frac{1}{2}x_2, \ x_7, \ x_4, \ x_9, \ x_6, \ \ldots)$$

for every $(x_n)_{n\in\mathbb{N}} \in l^1$.

The adjoint $S : l^\infty \to l^\infty$ of $T$ acts as follows: if $\gamma \in l^\infty$, $\gamma = (u_n)_{n\in\mathbb{N}}$, then $S\gamma = (v_n)_{n\in\mathbb{N}}$ where

$$v_n = \begin{cases} u_1 & \text{if } n = 1 \text{ or } n = 3 \\ \frac{1}{2}u_1 + \frac{1}{2}u_4 & \text{if } n = 2 \\ u_{2k+2} & \text{if } n = 2k, \ k \geq 2 \\ u_{2k-1} & \text{if } n = 2k + 1, \ k \geq 2 \end{cases};$$

that is,

$$S((u_n)_{n\in\mathbb{N}}) = (u_1, \ \frac{1}{2}u_1 + \frac{1}{2}u_4, \ u_1, \ u_6, \ u_3, \ u_8, \ u_5, \ u_{10}, \ \ldots)$$

for every $(u_n)_{n\in\mathbb{N}} \in l^\infty$.

In view of Example 1.1.9, the pair $(S, T)$ is Markov–Feller. It is easy to see that $\mathcal{D} = \{2k \mid k \in \mathbb{N},\ k \geq 2\}$, $\Gamma_c = \{2k - 1 \mid k \in \mathbb{N}\} \cup \{2\}$, and $\Gamma_{cp} = \{2k - 1 \mid k \in \mathbb{N}\}$. Note that $\varepsilon_x = \delta_1$ (the Dirac measure concentrated at 1) whenever $x \in \Gamma_{cp}$, and $\varepsilon_2 = \frac{1}{2}\delta_1$.                                      ∎

If $(S, T)$ is a uniquely ergodic Markov–Feller pair defined on a compact metric space, then $X = \Gamma_{cp}$ (this follows from a well-known result; see, for example, Proposition 1.2, p. 178 of Krengel's book [32]). However, even if $(S, T)$ is not uniquely ergodic, it may happen that $X = \Gamma_{cp}$; an example is provided by the rational rotations of the unit circle: if $X = \mathbb{R}/\mathbb{Z}$, if $a \in \mathbb{R}/\mathbb{Z}$ is such that (the class) $a$ contains a nonzero rational number, and if $(S_a, T_a)$ is the Markov–Feller pair of Example 1.1.11 (Rotations of the Unit Circle), then it is easy to see that $X = \Gamma_{cp}$, even though $(S_a, T_a)$ is not uniquely ergodic.

## 1.3   Special Topics: Topological Limits, Banach Limits, the Separability of $C_0(X)$, Order in Vector Spaces, and Equicontinuity

In Section 1.1 we went briefly over several aspects of the general theory of Markov–Feller operators, while in Section 1.2 we discussed various known types of and results on invariant probabilities of Markov–Feller operators. In this section we conclude the introductory chapter with an overview of several topics in general topology and functional analysis that will be used often in the book.

**Topological Limits.** As usual, let $X$ be a locally compact separable metric space, and let $(A_n)_{n \in \mathbb{N}}$ be a sequence of subsets of $X$. The *topological lower limit of the sequence* $(A_n)_{n \in \mathbb{N}}$ is a subset of $X$ (possibly empty) denoted $\underset{n \to \infty}{\mathrm{Li}}\ A_n$, and defined as follows:

$$\underset{n \to \infty}{\mathrm{Li}}\ A_n = \left\{ x \in X \ \middle| \ \begin{array}{l} \text{there exists a sequence } (x_n)_{n \in \mathbb{N}} \text{ of elements} \\ \text{of } X \text{ such that } x_n \in A_n \text{ for every } n \in \mathbb{N}, \text{ and} \\ (x_n)_{n \in \mathbb{N}} \text{ converges to } x \text{ in the metric topology of } X \end{array} \right\}.$$

The *topological upper limit of the sequence* $(A_n)_{n \in \mathbb{N}}$ is also a (possibly empty) subset of $X$ denoted $\underset{n \to \infty}{\mathrm{Ls}}\ A_n$, and defined by

$$\underset{n \to \infty}{\mathrm{Ls}}\ A_n = \left\{ x \in X \ \middle| \ \begin{array}{l} \text{there exists a subsequence } (A_{n_k})_{k \in \mathbb{N}} \text{ of } (A_n)_{n \in \mathbb{N}} \\ \text{and } x_k \in A_{n_k} \text{ for all } k \in \mathbb{N} \text{ such that the sequence} \\ (x_k)_{k \in \mathbb{N}} \text{ converges to } x \text{ in the metric topology of } X \end{array} \right\}.$$

We say that the sequence $(A_n)_{n \in \mathbb{N}}$ is *topologically convergent* if $\underset{n \to \infty}{\mathrm{Li}}\ A_n = \underset{n \to \infty}{\mathrm{Ls}}\ A_n$. In this case, we call the set $\underset{n \to \infty}{\mathrm{Li}}\ A_n = \underset{n \to \infty}{\mathrm{Ls}}\ A_n$ the *topological limit* of $(A_n)_{n \in \mathbb{N}}$. The topological limit of $(A_n)_{n \in \mathbb{N}}$ is denoted $\underset{n \to \infty}{\mathrm{Lt}}\ A_n$.

Note that $\underset{n\to\infty}{\text{Li}} A_n$, $\underset{n\to\infty}{\text{Ls}} A_n$, and $\underset{n\to\infty}{\text{Lt}} A_n$ (whenever $\underset{n\to\infty}{\text{Lt}} A_n$ exists, of course) are closed subsets of $X$. Note also that if $(A_n)_{n\in\mathbb{N}}$ is increasing (that is, $A_n \subseteq A_{n+1}$ for all $n \in \mathbb{N}$), then $\underset{n\to\infty}{\text{Lt}} A_n$ exists, and $\underset{n\to\infty}{\text{Lt}} A_n = \bigcup_{n=1}^{\infty} A_n$.

Details about topological limits can be found in the monograph by Kuratowski [34]. According to [34] the definitions of these limits emerged from the work of Painlevé. They have been used in the study of convolutions of measures (see the monographs by Heyer [28], and by Högnäs and Mukherjea [29]), and, recently, in the study of Markov–Feller operators by Lasota and Myjak [37], [38], [39], [40], and [41].

We will now discuss an application of the topological lower limits in describing the support of attractive probabilities of Markov–Feller pairs.

Let $(S, T)$ be a Markov–Feller pair defined on a locally compact separable metric space $(X, d)$. For every $x \in X$, and $n \in \mathbb{N} \cup \{0\}$ set $\sigma_n(x) = \text{supp}\,(T^n \delta_x)$; let $\sigma(x) = \underset{n\to\infty}{\text{Li}}\, \sigma_n(x)$ for every $x \in X$, and set $\sigma = \bigcap_{x\in X} \sigma(x)$. The next theorem appears in [72].

**Theorem 1.3.1.** *If the Markov–Feller pair $(S,\ T)$ has an attractive probability measure $\mu^*$, then $\text{supp}\ \mu^* = \sigma$.*

The approach taken in [72] has had a strong influence on our study of the supports of various types of invariant probabilities that is carried out in this work. Even though we will avoid using topological limits explicitly in stating and proving our results, from time to time we will pause to point out how we did arrive at the results by thinking in terms of topological limits.

**Banach Limits.** A *Banach limit* (or *Banach–Mazur limit*, or *generalized limit*) is a positive linear functional $L : l^\infty \to \mathbb{R}$ such that:

(1) $L(x_1,\ x_2,\ x_3,\ \dots\ ) = L(x_2,\ x_3,\ x_4,\ \dots\ )$ for every $(x_n)_{n\in\mathbb{N}} \in l^\infty$

and

(2) $L(1,\ 1,\ 1,\ \dots\ ) = 1$.

(The positivity of $L$ means, of course, that $L((x_n)_{n\in\mathbb{N}}) \geq 0$ whenever $x_n \geq 0$ for all $n \in \mathbb{N}$.)

Using the Hahn–Banach theorem it can be shown that Banach limits exist. We will need the following theorem (see Sucheston [65]):

**Theorem 1.3.2.** *Let $(x_n)_{n\in\mathbb{N}} \in l^\infty$. Then the sequences $\left( \underset{j\in\mathbb{N}\cup\{0\}}{\sup} \dfrac{1}{n} \sum_{i=0}^{n-1} x_{i+j} \right)_{n\in\mathbb{N}}$ and $\left( \underset{j\in\mathbb{N}\cup\{0\}}{\inf} \dfrac{1}{n} \sum_{i=0}^{n-1} x_{i+j} \right)_{n\in\mathbb{N}}$ converge and there exist Banach limits $L_M$ and $L_m$ such that*

$$L_M((x_n)_{n\in\mathbb{N}}) = \lim_{n\to\infty} \underset{j\in\mathbb{N}\cup\{0\}}{\sup} \frac{1}{n} \sum_{i=0}^{n-1} x_{i+j}$$

*and*

$$L_m((x_n)_{n\in\mathbb{N}}) = \lim_{n\to\infty} \inf_{j\in\mathbb{N}\cup\{0\}} \frac{1}{n}\sum_{i=0}^{n-1} x_{i+j}.$$

*Moreover, for every Banach limit $L$ we have*

$$L_m((x_n)_{n\in\mathbb{N}}) \leq L((x_n)_{n\in\mathbb{N}}) \leq L_M((x_n)_{n\in\mathbb{N}}).$$

*Consequently, there exists $s \in \mathbb{R}$ such that $L((x_n)_{n\in\mathbb{N}}) = s$ for every Banach limit $L$ if and only if the sequences $\left(\frac{1}{n}\sum_{i=0}^{n-1} x_{i+j}\right)_{n\in\mathbb{N}}$, $j \in \mathbb{N}\cup\{0\}$, converge to $s$ uniformly in $j$ (that is, for every $\varepsilon \in \mathbb{R}$, $\varepsilon > 0$ there exists $n_\varepsilon \in \mathbb{N}$ such that $\left|\frac{1}{n}\sum_{i=0}^{n-1} x_{i+j} - s\right| < \varepsilon$ for every $n \geq n_\varepsilon$ and $j \in \mathbb{N}\cup\{0\}$); in particular, if $(x_n)_{n\in\mathbb{N}}$ is a convergent sequence, then $L((x_n)_{n\in\mathbb{N}}) = \lim_{n\to\infty} x_n$ for every Banach limit $L$.*

Note that if $(x_n)_{n\in\mathbb{N}} \in l^\infty$ and $L$ is a Banach limit, then

$$\liminf_{n\to\infty} x_n \leq L((x_n)_{n\in\mathbb{N}}) \leq \limsup_{n\to\infty} x_n.$$

The results on Banach limits that we have described here can be found in the books by Dunford and Schwartz [15], Royden [61], and, of course, Sucheston's paper [65] where one can find a proof of Theorem 1.3.2.

### The Separability of $C_0(X)$ and Some Consequences

**Theorem 1.3.3.** *If $(X, d)$ is a locally compact separable metric space, then $C_0(X)$ is a separable Banach space.*

*Proof.* Let $D$ be a countable dense subset of $X$ (such a subset of $X$ exists because we assume that $X$ is separable).

For every $z \in X$ and $a \in \mathbb{R}$, $a > 0$ let $f_{z,a} : X \to \mathbb{R}$ be defined by $f_{z,a}(x) = d(x, X \setminus B(z, a))$ for every $x \in X$.

Set $\mathcal{D} = \{f_{z,a} \mid z \in D, a \in \mathbb{Q}, a > 0$ and $\overline{B(z,a)}$ is a compact subset of $X\}$ and let $\mathcal{A}$ be the subalgebra of $C_0(X)$ generated by $\mathcal{D}$.

Note that $\mathcal{D} \subseteq C_0(X)$ since each element of $\mathcal{D}$ is a continuous function with compact support (this is so since $|f_{z,a}(x) - f_{z,a}(y)| \leq d(x, y)$ for every $x \in X$, $y \in X$, $z \in X$, and $a \in \mathbb{R}$, $a > 0$; therefore, $f_{z,a}$ is continuous; obviously, $f_{z,a}$ has compact support because supp $f_{z,a} = \overline{B(z,a)}$).

Note also that $f \in \mathcal{A}$ if and only if there exist $n \in \mathbb{N}$, $f_j \in \mathcal{D}$, $j = 1, 2, 3, \ldots, n$ and $\alpha_{i_1, i_2, i_3, \ldots, i_n} \in \mathbb{R}$ for every $i_j \in \{0, 1, 2, \ldots, n\}$,

$j = 0, 1, 2, \ldots, n$, such that

$$f = \sum_{\substack{i_1=0 \\ i_2=0 \\ i_3=0 \\ \vdots \\ i_n=0}}^{n} \alpha_{i_1,\, i_2,\, i_3,\, \ldots,\, i_n} f_1^{i_1} f_2^{i_2} f_3^{i_3} \cdots f_n^{i_n} \tag{1.3.1}$$

(that is, $f \in \mathcal{A}$ if and only if $f$ is a "polynomial" with real coefficients, and with elements of $\mathcal{D}$ as "variables").

Let $\mathcal{A}_{\mathbb{Q}}$ be the set of all $f \in \mathcal{A}$ for which there exists a representation of the form (1.3.1) with rational coefficients (all the coefficients $\alpha_{i_1,\, i_2,\, i_3,\, \ldots,\, i_n}$ belong to $\mathbb{Q}$).

It is easy to see that $\mathcal{A}_{\mathbb{Q}}$ is a countable set, and that $\mathcal{A}_{\mathbb{Q}}$ is dense in $\mathcal{A}$ with respect to the topology on $\mathcal{A}$ induced by the uniform norm of $C_0(X)$. Thus, in order to prove that $C_0(X)$ is separable, it is enough to prove that $\mathcal{A}$ is dense in $C_0(X)$. In view of a version of the Stone–Weierstrass theorem (see Theorem D.23, p. 346 of Cohn's book [8]) we have to prove that:

(1) For every $x \in X$ there exists $f \in \mathcal{A}$ such that $f(x) \neq 0$.

(2) $\mathcal{A}$ separates the points of $X$ in the sense that for every $x, y \in X$ there exists $f \in \mathcal{A}$ such that $f(x) \neq f(y)$.

(1) Obviously, it is enough to prove that for every $x \in X$ there exists $f_{z,a} \in \mathcal{D}$ such that $f_{z,a}(x) > 0$.

Let $\alpha_z$, $z \in D$ be the numbers defined in the proof of Proposition 1.1.3. It was shown there that $X = \bigcup_{z \in D} B\left(z, \dfrac{\alpha_z}{2}\right)$; accordingly, $X = \bigcup_{z \in D} B\left(z, \dfrac{2\alpha_z}{3}\right)$. Thus, for every $x \in X$ there exists $z \in D$ such that $f_{z,\frac{2\alpha_z}{3}}(x) > 0$.

(2) In order to show that $\mathcal{A}$ separates the points of $X$, we will prove a bit more: we will show that for every $x, y \in X$, $x \neq y$ there exist $z \in D$, $a \in \mathbb{Q}$, $a > 0$ such that $f_{z,a} \in \mathcal{D}$, $f_{z,a}(x) > 0$, while $f_{z,a}(y) = 0$.

To this end, let $x, y \in X$, $x \neq y$. Since $D$ is dense in $X$, there exists a sequence $(z_n)_{n \in \mathbb{N}}$ of elements of $D$ that converges to $x$.

Using the proof of Proposition 1.1.3 we obtain that $\inf_{n \in \mathbb{N}} \alpha_{z_n} > 0$. Let $b = \min\left\{\dfrac{r}{2}, \inf_{n \in \mathbb{N}} \alpha_{z_n}\right\}$, let $n_0 \in \mathbb{N}$ be such that $d(x, z_{n_0}) < b$, and let $a \in \mathbb{Q}$ be such that $d(x, z_{n_0}) < a < b$. Then $\overline{B\left(z_{n_0}, a\right)}$ is a compact subset of $X$ and $x \in B\left(z_{n_0}, a\right)$; accordingly, $f_{z_{n_0},a} \in \mathcal{D}$ and $f_{z_{n_0},a}(x) > 0$. Since

$$d(z_{n_0}, y) \geq d(x, y) - d(z_{n_0}, x) > r - b \geq \frac{r}{2} > a,$$

it follows that $f_{z_{n_0},a}(y) = 0$.

Since the assertions (1) and (2) are true, it follows that $\mathcal{A}$ is dense in $C_0(X)$.
□

Let $(Y, d_1)$ and $(Z, d_2)$ be two metric spaces. We say that $f : Y \to Z$ is a *Lipschitz function* if there exists $\lambda \in \mathbb{R}$, $\lambda \geq 0$ such that $d_2(f(x), f(y)) \leq \lambda d_1(x, y)$ for every $x, y \in Y$.

If $(X, d)$ is our usual locally compact separable metric space, set $\mathcal{L}ip(X) = \{f \in C_0(X) \mid f \text{ is a Lipschitz function}\}$.

We conclude this subsection with a corollary of Theorem 1.3.3.

**Corollary 1.3.4.** *If $(X, d)$ is a locally compact separable metric space, then the vector space of all Lipschitz functions with compact support is dense in $C_c(X)$ and in $C_0(X)$ (with respect to the topology generated by the uniform norm, of course). Consequently, $\mathcal{L}ip(X)$ is dense in $C_0(X)$.*

*Proof.* Let $\mathcal{A}$ be the subalgebra constructed in the proof of Theorem 1.3.3. Since the sum or the product of two Lipschitz functions is Lipschitz, it follows that all the functions in $\mathcal{A}$ have compact supports and are Lipschitz. Since $\mathcal{A}$ is dense in $C_0(X)$, it follows that the assertions of the corollary are true.                    □

**Vector Lattices, Banach Lattices, and Positive Operators.** Our goal in this subsection is to review briefly a few notions and results of the theory of vector lattices and the theory of positive operators on vector lattices because these notions and results are useful in obtaining a better understanding of many of the topics discussed in this volume. The material presented in this subsection can be found in much greater detail in almost every monograph or textbook that deals with vector lattices and positive operators (for example, see the books by Abramovich and Aliprantis [1], Aliprantis and Burkinshaw [2], Schaefer [63], and Zaanen [71]).

Let $E$ be a real vector space (all the vector spaces that appear in this volume are real vector spaces), and assume that $E$ is endowed with an order relation $\leq$. The ordered pair $(E, \leq)$ is called an *ordered vector space* if the following conditions are satisfied:

- if $x \in E$ and $y \in E$ are such that $x \leq y$, then $x + z \leq y + z$ whenever $z \in E$;

- if $x \in E$ and $y \in E$ are such that $x \leq y$, then $\lambda x \leq \lambda y$ whenever $\lambda \in \mathbb{R}$, $\lambda \geq 0$.

In general, the order relation under consideration will be clear from the context, so we will use the notation $E$ rather than $(E, \leq)$ whenever $(E, \leq)$ is an ordered vector space.

If $E$ is an ordered vector space, and if $u \in E$, then we say that $u$ is a *positive element of $E$* if $0 \leq u$.

An ordered vector space $E$ is called a *Riesz space* (or a *vector lattice*, or a *linear lattice*) if $\sup\{x, y\}$ and $\inf\{x, y\}$ exist whenever $x \in E$ and $y \in E$. We will use the notations $x \vee y = \sup\{x, y\}$ and $x \wedge y = \inf\{x, y\}$, $x \in E$, $y \in E$.

Let $E$ be a Riesz space. It is customary to use the following notations: $x^+ = x \vee 0$, $x^- = (-x) \vee 0$, and $|x| = x \vee (-x)$ whenever $x \in E$.

Let $E$ be a Riesz space, and assume that $E$ is endowed with a norm $\| \ \|$ that defines a Banach space structure on $E$. We say that $E$ is a *Banach lattice* if the following condition is satisfied: if $x \in E$ and $y \in E$ are such that $|x| \leq |y|$, then $\|x\| \leq \|y\|$.

Most Banach spaces that we encounter in our everyday mathematical life are Banach lattices when endowed with their standard norms and order relations. Thus, the spaces $C_0(X)$, $C_b(X)$, $B_b(X)$ are all Banach lattices when endowed with the pointwise order (recall that the pointwise order is defined as follows: $f \leq g$ if, by definition, $f(t) \leq g(t)$ for every $t \in X$ whenever $f$ and $g$ belong to the space under consideration), where $X$ is our usual locally compact separable metric space; the space $\mathcal{M}(X)$, where $X$ is as above, is a Banach lattice when endowed with the standard order relation defined by $\mu \leq \nu$ if, by definition, $\mu(A) \leq \nu(A)$ for every $A \in \mathcal{B}(X)$; let $(Y, \Sigma, \mu)$ be a measure space, let $p \in \mathbb{R} \cup \{+\infty\}$, $1 \leq p \leq +\infty$, and let $L^p(Y, \Sigma, \mu)$ be the usual $L^p$-space; then $L^p(Y, \Sigma, \mu)$ is a Banach lattice with respect to its standard order relation defined as follows: $\bar{f}_1 \leq \bar{f}_2$ if, by definition, there exist two measurable functions $g_1, g_2$ such that $g_i$ is in the class $\bar{f}_i$, $i = 1, 2$, and such that $g_1 \leq g_2$ $\mu$-a.e. where $\bar{f}_i \in L^p(Y, \Sigma, \mu)$, $i = 1, 2$.

Of course, there exist Riesz spaces that are not Banach lattices. For example, consider the vector space $\mathbb{R}^2$ endowed with the lexicographical order (recall that the *lexicographical order* $\leq_L$ is defined as follows: $(a, b) \leq_L (x, y)$ if either $a < x$, or else $a = x$ and $b \leq y$, $(a, b) \in \mathbb{R}^2$, $(x, y) \in \mathbb{R}^2$). It can be shown that $\mathbb{R}^2$ endowed with the lexicographical order is a Riesz space, and that no norm on $\mathbb{R}^2$ can define a Banach lattice structure on $\mathbb{R}^2$ under the lexicographical order.

A Banach lattice $E$ is called an *AL-space* (*abstract $L^1$-space*) if its norm is additive; that is, if $\|x + y\| = \|x\| + \|y\|$ whenever $x \in E$, $x \geq 0$, and $y \in E$, $y \geq 0$. The space $\mathcal{M}(X)$ defined in Section 1.1, and the space $L^1(Y, \Sigma, \mu)$ where $(Y, \Sigma, \mu)$ is a measure space are AL-spaces; the spaces $C_0(X)$, $C_b(X)$, and $B_b(X)$ that we defined in Section 1.1 are examples of Banach lattices that are not AL-spaces (except, of course, for the trivial case when $X$ has only one element since in this case $C_0(X)$, $C_b(X)$, and $B_b(X)$ can all be identified with $\mathbb{R}$).

A Banach lattice $E$ is called an *AM-space* if $\|x \vee y\| = \max\{\|x\|, \|y\|\}$ whenever $x \in E$ and $y \in E$ are such that $x \wedge y = 0$. The spaces $C_0(X)$, $C_b(X)$, and $B_b(X)$ that we mentioned above are AM-spaces; if $X$ has at least two elements, the space $\mathcal{M}(X)$ is not an AM-space. If $(Y, \Sigma, \mu)$ is a measure space such that the measure $\mu$ has at least two distinct nonzero real values, then the spaces $L^p(Y, \Sigma, \mu)$, $1 < p < +\infty$ are examples of Banach lattices which are neither AL-spaces, nor AM-spaces ($L^\infty(Y, \Sigma, \mu)$ is an AM-space).

An AM-space $E$ is called an *AM-space with unit* if the closed unit ball $\{u \in E \mid \|u\| \leq 1\}$ has a largest element. If the closed unit ball of $E$ has a largest element, then this element is unique, and is called a *(strong order) unit* of $E$. For example, the spaces $C_b(X)$ and $B_b(X)$ are AM-spaces with unit (the unit in both spaces is $1_X$); by contrast, $C_0(X)$ is an example of an AM-space without unit whenever $(X, d)$ is not compact.

Let $E$ and $F$ be two Riesz spaces. A linear operator $T : E \to F$ is called a *positive operator* if $Tu \geq 0$ whenever $u \in E$, $u \geq 0$. It can be shown (see, for example, Theorem 5.3, p. 84 of Schaefer [63]) that if $E$ and $F$ are Banach lattices, and $T$ is a positive operator, then $T$ is bounded (continuous); in particular, a positive linear functional $w$ on a Banach lattice $E$ (that is, a positive linear operator $w : E \to \mathbb{R}$) is bounded.

Let $E$ be a Banach lattice, and let $E'$ be the (topological) dual of $E$ (that is, $E'$ is the Banach space of all linear bounded (real-valued) functionals on $E$). We say that $u \in E'$ is a *positive element of $E'$* if $u$ thought of as a linear operator with domain $E$ and codomain $\mathbb{R}$ is a positive operator. On $E'$ we can define an order relation $\leq$ as follows: $u \leq v$ if, by definition $v - u$ is a positive element of $E'$, $u \in E'$, $v \in E'$. It can be shown that the order relation $\leq$ on these functionals defines a Banach lattice structure on $E'$.

Let $E$ and $F$ be two Banach lattices. We say that $E$ is *isometric and order isomorphic* to $F$ if there exists a positive linear operator $T : E \to F$ such that $T$ is one-to-one, onto, an isometry, and such that the inverse $T^{-1}$ of $T$ is again a positive operator. In this case we often think of $E$ and $F$ as being the same space, and we identify each $u \in E$ with the element $Tu$ of $F$. For example, the dual $(C_0(X))'$ of $C_0(X)$ is order isomorphic and isometric to $\mathcal{M}(X)$ in our usual setting of the book (that is, whenever $X$ is a locally compact separable metric space), so we think of the elements of $(C_0(X))'$ as being elements of $\mathcal{M}(X)$.

As usual, if $E$ and $F$ are Banach spaces, and $T : E \to F$ is a linear operator, we say that $T$ is a *contraction* if $T$ is bounded and $\|T\| \leq 1$.

Let $E$ be an AL-space. A linear operator $T : E \to E$ is called a *Markov operator* if $Tu \geq 0$ and $\|Tu\| = \|u\|$ for every $u \in E$, $u \geq 0$. Clearly, a Markov operator is positive. Moreover, if $T : E \to E$ is a Markov operator, then $\|T\| = 1$ (so, $T$ is a contraction) because the positivity of $T$ implies that

$$\|T\| = \sup\{\|Tu\| \mid u \in E,\ u \geq 0,\ \|u\| \leq 1\}.$$

Note that the notion of Markov operator defined here is a natural extension of the notions of Markov operator that were introduced in Section 1.1 and in the subsection *Almost Everywhere Convergence Results* of Section 1.2 in the sense that the Markov operators discussed in Section 1.1 and in the above-mentioned subsection of Section 1.2 are particular cases of Markov operators as defined here.

We will conclude this subsection by discussing another proof of Lemma 1.2.3. In order to offer the proof, we need some preparation.

Let $E$ be a Riesz space.

A vector subspace $V$ of $E$ is called an *(order) ideal* of $E$ if the following two conditions are satisfied:

(i) If $u \in V$, then $|u| \in V$.

(ii) If $u \in V$, $v \in E$, and $0 \leq v \leq u$, then $v \in V$.

A vector subspace $W$ of $E$ is called a *band of* (or *in*) $E$ if $W$ is an ideal, and if for every subset $A$ of $W$ such that $\sup A$ exists in $E$, it follows that $\sup A$ is an element of $W$.

The intersection of a family of bands of $E$ is again a band in $E$. Since $E$ is obviously a band of itself, it follows that given a nonempty subset $A$ of $E$ there exists a smallest band of $E$ that contains $A$. This smallest band is denoted by $B(A)$ and is called the *band generated by* $A$.

Let $I$ be an ideal of $E$, and let $B(I)$ be the band generated by $I$. It can be shown (see, for example, Theorem 7.8, pp. 32–33 of Zaanen [71]) that if $h \in B(I)$, $h \geq 0$, then

$$h = \sup\{u \in I \mid 0 \leq u \leq h\}. \tag{1.3.2}$$

Let $A$ be a nonempty subset of $E$, and set

$$A^d = \{u \in E \mid |u| \wedge |v| = 0 \text{ for every } v \in A\}.$$

Also, let $A^{dd} = (A^d)^d$. The sets $A^d$ and $A^{dd}$ are called the *disjoint complement* and the *second disjoint complement of* $A$, respectively. It is easy to see that $A \subseteq A^{dd}$, and it can be shown (see, for example, Theorem 8.4, pp. 36–37 of Zaanen [71]) that both $A^d$ and $A^{dd}$ are bands of $E$.

If $I$ is an ideal of $E$, then $I \subseteq I^{dd}$ and $I^{dd}$ is a band; thus, a natural question (that is also relevant to our discussion) is: for what kind of Riesz spaces $E$ it is true that $I^{dd} = B(I)$ for every ideal $I$ of $E$? (Note that, in general, it is not true that $B(I) = I^{dd}$ for an ideal $I$ of a Riesz space $E$; indeed, consider the Riesz space $\mathbb{R}^2$ endowed with the lexicographical order that we discussed earlier in this subsection, and set $I = \{(x, y) \in \mathbb{R}^2 \mid x = y\}$; it is not difficult to see that $I$ is a band of $\mathbb{R}^2$ (therefore, I is an ideal, and $I = B(I)$), and that $I^{dd} = \mathbb{R}^2$.)

A Riesz space $E$ is said to be *Archimedean* if $\inf\left\{\dfrac{1}{n}u \mid n \in \mathbb{N}\right\} = 0$ for every $u \in E$, $u \geq 0$. It can be shown (see, for example, Proposition 5.2, pp. 83–84 of Schaefer [63]) that every Banach lattice is Archimedean. The Riesz space $\mathbb{R}^2$ endowed with the lexicographic order is an example of a vector lattice that fails to be Archimedean. The Archimedean Riesz spaces can be used to answer completely the question of the previous paragraph; that is, using Theorem 9.6, p. 45 of Zaanen [71] and Exercise 9.8, p. 46 of [71], we obtain that a Riesz space $E$ is Archimedean if and only if $I^{dd} = B(I)$ for every ideal $I$ of $E$.

Our discussion on ideals, bands, and disjoint complements makes it possible now for us to discuss another proof of Lemma 1.2.3 that we mentioned earlier.

*Second Proof of Lemma* 1.2.3. To begin with, we note that for every $x_0 \in X$ there exists $f \in C_0(X)$ (actually, we can define $f$ to have compact support) such that $f \geq 0$ and $f(x_0) > 0$ (indeed, since $X$ is locally compact, there exists $r \in \mathbb{R}$, $r > 0$ such that $\overline{B(x_0, r)}$ is a compact subset of $X$; Let $f : X \to \mathbb{R}$ be defined by $f(x) = d(x, X \setminus B(x_0, r))$ for every $x \in X$; then $f$ is continuous, $f(x_0) \geq r > 0$, and $f$ has compact support (since $\operatorname{supp} f = \overline{B(x_0, r)}$)).

It follows that the disjoint complement $(C_0(X))^d$ of $C_0(X)$ in $C_b(X)$ is $\{0\}$. Therefore, $(C_0(X))^{dd} = C_b(X)$. Since $C_b(X)$ is a Banach lattice, it follows that $C_b(X)$ is an Archimedean Riesz space. Since $C_0(X)$ is an ideal in $C_b(X)$, it follows that the band generated by $C_0(X)$ is actually $C_b(X)$. Using the inequality (1.3.2), we note that the assertion of the lemma is true. □

Even though the above proof of Lemma 1.2.3 uses more sophisticated tools than the first proof, it has the advantage that it suggests extensions of the lemma; the extensions could then be used to generalize some of the results discussed in this volume. An example of such an extension is the following proposition:

**Proposition 1.3.5 (Extension of Lemma 1.2.3).** *Assume that $(X, d)$ is just a metric space (not necessarily locally compact or separable), and let $C_b(X)$ be the Banach lattice of all real-valued continuous bounded functions defined on $X$ (we assume, as usual in this book, that $C_b(X)$ is endowed with the uniform (sup) norm and the pointwise order). Let $I$ be an ideal in $C_b(X)$, and assume that $I$ has the property that for every $x \in X$ there exists $u \in I$, $u \geq 0$, such that $u(x) > 0$. Then $h = \sup\{u \in I \,|\, 0 \leq u \leq h\}$ for every $h \in C_b(X)$, $h \geq 0$.*

*Proof.* The proof follows along the lines of the second proof of Lemma 1.2.3. Indeed, if $I$ is an ideal of $C_b(X)$ that satisfies the condition of the proposition, then $I^d = \{0\}$, and $I^{dd} = C_b(X)$. Therefore, we can use equality (1.3.2) in order to complete the proof. □

**Equicontinuity.** As we do most of the time in this book, throughout this subsection we assume given a locally compact separable metric space $(X, d)$.

Let $A$ be a nonempty subset of $X$, and let $\mathcal{F}$ be a family of real-valued functions defined on $X$. We say that $\mathcal{F}$ is *equicontinuous on $A$* if for every convergent sequence $(x_k)_{k \in \mathbb{N}}$ such that the limit $x$ of $(x_k)_{k \in \mathbb{N}}$ is an element of $A$, and for every $\varepsilon \in \mathbb{R}$, $\varepsilon > 0$ there exists $k_\varepsilon \in \mathbb{N}$ such that $|f(x_k) - f(x)| < \varepsilon$ for every $k \geq k_\varepsilon$ and $f \in \mathcal{F}$; we say that $\mathcal{F}$ is *uniformly equicontinuous on $A$* if for every $\varepsilon \in \mathbb{R}$, $\varepsilon > 0$ there exists $\delta \in \mathbb{R}$, $\delta > 0$, such that $|f(x) - f(y)| < \varepsilon$ for every $f \in \mathcal{F}$, and $x \in A$, $y \in A$, such that $d(x, y) < \delta$. If $(f_n)_{n \in \mathbb{N}}$ is a sequence of real-valued functions on $X$, we say that $(f_n)_{n \in \mathbb{N}}$ is *equicontinuous* or *uniformly equicontinuous on $A$* if the range $\{f_n | n \in \mathbb{N}\}$ of the sequence $(f_n)_{n \in \mathbb{N}}$ is equicontinuous or uniformly equicontinuous on $A$, respectively. If a family $\mathcal{F}$ is equicontinuous or uniformly equicontinuous on $X$, we simply say that $\mathcal{F}$ is *equicontinuous* or *uniformly equicontinuous*, respectively. Similarly, if the range $\{f_n | n \in \mathbb{N}\}$ of a sequence $(f_n)_{n \in \mathbb{N}}$ is equicontinuous or uniformly equicontinuous, we say that the sequence itself is *equicontinuous* or *uniformly equicontinuous*, respectively.

**Lemma 1.3.6.** *Let $(f_n)_{n \in \mathbb{N}}$ be a sequence of real-valued functions defined on $X$, and let $K$ be a compact nonempty subset of $X$. Then $(f_n)_{n \in \mathbb{N}}$ is equicontinuous on $K$ if and only if $(f_n)_{n \in \mathbb{N}}$ is uniformly equicontinuous on $K$.*

*Proof.* Clearly, if $(f_n)_{n \in \mathbb{N}}$ is uniformly equicontinuous on $K$, then $(f_n)_{n \in \mathbb{N}}$ is also equicontinuous on $K$.

Thus, we only have to prove that if $(f_n)_{n\in\mathbb{N}}$ is equicontinuous on $K$, then the sequence is also uniformly equicontinuous on $K$. To this end, assume that $(f_n)_{n\in\mathbb{N}}$ is not uniformly equicontinuous on $K$. Then there exists $\varepsilon_0 \in \mathbb{R}$, $\varepsilon_0 > 0$, such that for every $k \in \mathbb{N}$ there exist $x_k \in K$, $y_k \in K$, and $n_k \in \mathbb{N}$ such that $d(x_k, y_k) < \dfrac{1}{k}$, but $|f_{n_k}(x_k) - f_{n_k}(y_k)| \geq \varepsilon_0$. Since $K$ is compact, there exists a convergent subsequence $(x_{k_l})_{l\in\mathbb{N}}$ of $(x_k)_{k\in\mathbb{N}}$. Let $x^* = \lim\limits_{l\to\infty} x_{k_l}$. If we assume that $(f_n)_{n\in\mathbb{N}}$ is equicontinuous, then taking into consideration that both $(x_{k_l})_{l\in\mathbb{N}}$ and $(y_{k_l})_{l\in\mathbb{N}}$ converge to $x^*$, and that $x^* \in K$, we obtain that there exists $l_0 \in \mathbb{N}$ such that $\left|f_n\left(x_{k_{l_0}}\right) - f_n(x^*)\right| < \dfrac{\varepsilon_0}{2}$ and $\left|f_n\left(y_{k_{l_0}}\right) - f_n(x^*)\right| < \dfrac{\varepsilon_0}{2}$ for every $n \in \mathbb{N}$. In particular, for $n = n_{k_{l_0}}$ it follows that

$$\left|f_{n_{k_{l_0}}}\left(x_{k_{l_0}}\right) - f_{n_{k_{l_0}}}\left(y_{k_{l_0}}\right)\right| \leq \left|f_{n_{k_{l_0}}}\left(x_{k_{l_0}}\right) - f_{n_{k_{l_0}}}(x^*)\right|$$

$$+ \left|f_{n_{k_{l_0}}}(x^*) - f_{n_{k_{l_0}}}\left(y_{k_{l_0}}\right)\right| < \frac{\varepsilon_0}{2} + \frac{\varepsilon_0}{2} = \varepsilon_0,$$

so we have obtained a contradiction which stems from the assumption that $(f_n)_{n\in\mathbb{N}}$ is equicontinuous but not uniformly equicontinuous on $K$. $\qquad\square$

As one may expect, we say that a family or a sequence of real-valued functions defined on $X$ is *(uniformly) equicontinuous on the compact subsets of $X$* if the family or the sequence is (uniformly) equicontinuous on every nonempty compact subset of $X$.

**Proposition 1.3.7.** *Let $(f_n)_{n\in\mathbb{N}}$ be a sequence of real-valued functions defined on $X$. The following assertions are equivalent:*

(a) *$(f_n)_{n\in\mathbb{N}}$ is equicontinuous (on $X$).*

(b) *$(f_n)_{n\in\mathbb{N}}$ is equicontinuous on the compact subsets of $X$.*

(c) *$(f_n)_{n\in\mathbb{N}}$ is uniformly equicontinuous on the compact subsets of $X$.*

*Proof.* (a) $\Rightarrow$ (b) is obvious. (b) and (c) are equivalent by Lemma 1.3.5.

(b) $\Rightarrow$ (a): Let $(x_k)_{k\in\mathbb{N}}$ be a convergent sequence of elements of $X$, and set $x = \lim\limits_{k\to\infty} x_k$. Then the set $K = \{x_k | k \in \mathbb{N}\} \cup \{x\}$ is obviously compact in $X$. Since we assume that $(f_n)_{n\in\mathbb{N}}$ is equicontinuous on $K$, since $(x_k)_{k\in\mathbb{N}}$ is a convergent sequence of elements of $K$, and since $x \in K$, it is straightforward that for every $\varepsilon \in \mathbb{R}$, $\varepsilon > 0$ there exists $k_\varepsilon \in \mathbb{N}$ such that $|f_n(x_k) - f_n(x)| < \varepsilon$ for every $k \geq k_\varepsilon$ and $n \in \mathbb{N}$. $\qquad\square$

Note that Lemma 1.3.6 and Proposition 1.3.7 remain valid when stated for families rather than sequences of real-valued functions on $X$. Indeed, all the arguments used in the proofs of Lemma 1.3.6 and Proposition 1.3.7 can be easily adapted for the case in which we deal with a family rather than a sequence of functions. The only place where we might need an explanation is in the proof of the fact that if a family $\mathcal{F}$ is equicontinuous on a nonempty compact subset $K$ of $X$, then $\mathcal{F}$ is uniformly equicontinuous on $K$ (in the proof of the assertion of Lemma 1.3.6 for families): if we assume that $\mathcal{F}$ is not uniformly equicontinuous on $K$ (but $\mathcal{F}$ is equicontinuous on $K$), then there exists $\varepsilon_0 \in \mathbb{R}$, $\varepsilon_0 > 0$, such that for every $k \in \mathbb{N}$ there exist $f_k \in \mathcal{F}$, $x_k \in K$, $y_k \in K$ such that $d(x_k, y_k) < \dfrac{1}{k}$ and $|f_k(x_k) - f_k(y_k)| \geq \varepsilon_0$; but then the sequence $(f_k)_{k \in \mathbb{N}}$ (of elements of $\mathcal{F}$) is not uniformly continuous on $K$; by Lemma 1.3.6 (the assertion of the lemma for sequences), the sequence $(f_k)_{k \in \mathbb{N}}$ is not equicontinuous on $K$; since $(f_k)_{k \in \mathbb{N}}$ is not equicontinuous on $K$, it follows that $\mathcal{F}$ is not equicontinuous on $K$, so we have obtained a contradiction.

Let $(f_k)_{k \in \mathbb{N}}$ be a sequence of real-valued functions defined on $X$. We say that $(f_k)_{k \in \mathbb{N}}$ is a *uniformly Cauchy sequence on the compact subsets of $X$* if for every nonempty compact subset $K$ of $X$ and for every $\varepsilon \in \mathbb{R}$, $\varepsilon > 0$, there exists $k_{\varepsilon,K} \in \mathbb{N}$ such that $|f_k(x) - f_l(x)| < \varepsilon$ for every $k \geq k_{\varepsilon,K}$, $l \geq k_{\varepsilon,K}$, and $x \in K$. We say that $(f_k)_{k \in \mathbb{N}}$ *converges uniformly on the compact subsets of $X$* if there exists $f : X \to \mathbb{R}$ such that for every nonempty compact subset $K$ of $X$, and for every $\varepsilon \in \mathbb{R}$, $\varepsilon > 0$, there exists $k_{\varepsilon,K} \in \mathbb{N}$ such that $|f_k(x) - f(x)| < \varepsilon$ for every $k \geq k_{\varepsilon,K}$ and $x \in K$; in this case, we call $f$ the *uniform limit of $(f_k)_{k \in \mathbb{N}}$ on the compact subsets of $X$*, and we also say that $(f_k)_{k \in \mathbb{N}}$ *converges uniformly to $f$ on the compact subsets of $X$*.

**Proposition 1.3.8.** *Let $(f_k)_{k \in \mathbb{N}}$ be a sequence of real-valued functions defined on $X$. The following assertions are equivalent:*

(a) *The sequence $(f_k)_{k \in \mathbb{N}}$ converges uniformly on the compact subsets of $X$.*

(b) *The sequence $(f_k)_{k \in \mathbb{N}}$ is a uniformly Cauchy sequence on the compact subsets of $X$.*

*Proof.* (a) $\Rightarrow$ (b): Since we assume that (a) holds true, there exists $f : X \to \mathbb{R}$ such that $(f_k)_{k \in \mathbb{N}}$ converges uniformly to $f$ on the compact subsets of $X$.

Let $K$ be a compact subset of $X$, and let $\varepsilon \in \mathbb{R}$, $\varepsilon > 0$. Then there exists $k_\varepsilon \in \mathbb{N}$ such that $|f_k(x) - f(x)| < \dfrac{\varepsilon}{2}$ for every $k \geq k_\varepsilon$ and $x \in K$. Therefore,

$$|f_k(x) - f_l(x)| \leq |f_k(x) - f(x)| + |f(x) - f_l(x)| < \frac{\varepsilon}{2} + \frac{\varepsilon}{2} = \varepsilon \text{ for every } k \geq k_\varepsilon, l \geq$$

$k_\varepsilon$, and $x \in K$.

(b) $\Rightarrow$ (a): Since we assume that $(f_k)_{k \in \mathbb{N}}$ is uniformly Cauchy on the compact subsets of $X$, and since $\{x\}$ (the set that has only one element, namely, $x$) is a compact subset of $X$, it follows that $(f_k(x))_{k \in \mathbb{N}}$ is a convergent sequence of real numbers for every $x \in X$; thus, it makes sense to define $f : X \to \mathbb{R}$, $f(x) = \lim\limits_{k \to \infty} f_k(x)$ for every $x \in X$.

Now let $K$ be a compact subset of $X$, and let $\varepsilon \in \mathbb{R}$, $\varepsilon > 0$. Then there exists $k_\varepsilon \in \mathbb{N}$ such that $|f_k(x) - f_{k+l}(x)| < \dfrac{\varepsilon}{2}$ for every $k \geq k_\varepsilon$, $l \in \mathbb{N}$, and $x \in K$; since $\lim\limits_{l \to \infty} f_{k+l}(x) = f(x)$ for every $x \in K$, it follows that $|f_k(x) - f(x)| \leq \dfrac{\varepsilon}{2} < \varepsilon$ for every $k \geq k_\varepsilon$ and $x \in K$. Thus, $(f_k)_{k \in \mathbb{N}}$ converges uniformly to $f$ on the compact subsets of $X$.                                                                                □

As usual, we say that a sequence $(f_k)_{k \in \mathbb{N}}$ of real-valued functions defined on $X$ is *bounded* if there exists $M \in \mathbb{R}$, $M \geq 0$ such that $|f_k(x)| \leq M$ for every $k \in \mathbb{N}$ and $x \in X$.

**Proposition 1.3.9.** *Let $(f_k)_{k \in \mathbb{N}}$ be a bounded sequence of elements of $C_b(X)$, and assume that there exists $f : X \to \mathbb{R}$ such that $(f_k)_{k \in \mathbb{N}}$ converges uniformly to $f$ on the compact subsets of $X$. Then:*

(a) $f \in C_b(X)$.

(b) $(f_k)_{k \in \mathbb{N}}$ *is equicontinuous.*

*Proof.* (a) We first note that since $(f_k)_{k \in \mathbb{N}}$ converges uniformly to $f$ on the compact subsets of $X$, it follows that $(f_k)_{k \in \mathbb{N}}$ converges also pointwise to $f$ (that is, $(f_k(x))_{k \in \mathbb{N}}$ converges to $f(x)$ for every $x \in X$) as we already pointed out at the beginning of the proof of (b) $\Rightarrow$ (a) in the proof of Proposition 1.3.8. Since we assume that $(f_k)_{k \in \mathbb{N}}$ is a bounded sequence, it follows that $f$ is a bounded function.

We now prove that $f$ is continuous. Thus, we have to prove that the sequence $(f(x_l))_{l \in \mathbb{N}}$ converges to $f(x)$ for every convergent sequence $(x_l)_{l \in \mathbb{N}}$ of elements of $X$ such that $x = \lim\limits_{l \to \infty} x_l$.

To this end, let $(x_l)_{l \in \mathbb{N}}$ be a convergent sequence of elements of $X$, set $x = \lim\limits_{l \to \infty} x_l$, let $\varepsilon \in \mathbb{R}$, $\varepsilon > 0$, and set $K = \{x_l | l \in \mathbb{N}\} \cup \{x\}$. Since $(f_k)_{k \in \mathbb{N}}$ converges uniformly to $f$ on the compact subsets of $X$, and since $K$ is compact, it follows that there exists $k_\varepsilon \in \mathbb{N}$ such that $|f_k(x_l) - f(x_l)| < \dfrac{\varepsilon}{3}$ for every $k \geq k_\varepsilon$ and $l \in \mathbb{N}$, and such that $|f_k(x) - f(x)| < \dfrac{\varepsilon}{3}$ for every $k \geq k_\varepsilon$. In particular, $|f_{k_\varepsilon}(x_l) - f(x_l)| < \dfrac{\varepsilon}{3}$ for every $l \in \mathbb{N}$ and $|f_{k_\varepsilon}(x) - f(x)| < \dfrac{\varepsilon}{3}$. Since $f_{k_\varepsilon}$ is a continuous function, there exists $l_\varepsilon \in \mathbb{N}$ such that $|f_{k_\varepsilon}(x_l) - f_{k_\varepsilon}(x)| < \dfrac{\varepsilon}{3}$ for every $l \geq l_\varepsilon$. We obtain that

$$|f(x_l) - f(x)| \leq |f(x_l) - f_{k_\varepsilon}(x_l)| + |f_{k_\varepsilon}(x_l) - f_{k_\varepsilon}(x)| + |f_{k_\varepsilon}(x) - f(x)|$$

$$< \frac{\varepsilon}{3} + \frac{\varepsilon}{3} + \frac{\varepsilon}{3} = \varepsilon$$

for every $l \geq l_\varepsilon$. Thus, $(f(x_l))_{l \in \mathbb{N}}$ converges to $f(x)$.

(b) In order to prove that $(f_k)_{k \in \mathbb{N}}$ is equicontinuous, we have to prove that for every convergent sequence $(x_l)_{l \in \mathbb{N}}$ of elements of $X$, and for every $\varepsilon \in \mathbb{R}$, $\varepsilon > 0$, there exists $l_\varepsilon \in \mathbb{N}$ such that $|f_k(x_l) - f_k(x)| < \varepsilon$ for every $k \in \mathbb{N}$ and $l \geq l_\varepsilon$, provided that $x = \lim\limits_{l \to \infty} x_l$.

Thus, let $(x_l)_{l \in \mathbb{N}}$ be a convergent sequence of elements of $X$, let $x = \lim\limits_{l \to \infty} x_l$, and let $\varepsilon \in \mathbb{R}$, $\varepsilon > 0$. Since the subset $K = \{x_l | l \in \mathbb{N}\} \cup \{x\}$ of $X$ is compact, and since $(f_k)_{k \in \mathbb{N}}$ converges uniformly to $f$ on the compact subsets of $X$, there exists $k_\varepsilon \in \mathbb{N}$ such that $|f_k(y) - f(y)| < \dfrac{\varepsilon}{3}$ for every $k \geq k_\varepsilon$ and $y \in K$.

Since $f_1, f_2, \ldots, f_{k_\varepsilon - 1}$ are continuous functions, and since (by (a)) $f$ is also continuous, it follows that there exists $l_\varepsilon \in \mathbb{N}$ such that $|f_k(x_l) - f_k(x)| < \varepsilon$ for every $k = 1, 2, 3, \ldots, k_\varepsilon - 1$ and every $l \geq l_\varepsilon$, and such that $|f(x_l) - f(x)| < \dfrac{\varepsilon}{3}$ for every $l \geq l_\varepsilon$.

Now let $k \in \mathbb{N}$. If $k < k_\varepsilon$, then $|f_k(x_l) - f_k(x)| < \varepsilon$ for every $l \geq l_\varepsilon$; if $k \geq k_\varepsilon$, then

$$|f_k(x_l) - f_k(x)| \leq |f_k(x_l) - f(x_l)| + |f(x_l) - f(x)| + |f(x) - f_k(x)|$$

$$< \frac{\varepsilon}{3} + \frac{\varepsilon}{3} + \frac{\varepsilon}{3} = \varepsilon$$

for every $l \geq l_\varepsilon$. $\qquad\square$

# Chapter 2

# The Krylov–Bogolioubov–Beboutoff–Yosida (KBBY) Decomposition

The central goal of this chapter is to obtain "formulas" for the supports of the ergodic measures of a Markov-Feller operator (see Theorem 2.2.2 of Section 2.2). However, we need a framework to allow us to obtain such formulas. It turns out that such a framework can be obtained by extending to our setting an ergodic decomposition that has emerged from the works of Krylov and Bogolioubov [33], Beboutoff [5], and Yosida [68] and [69]. We call the decomposition the KBBY decomposition, and we obtain the extension in Section 2.1. The main ingredients for the extension are the Lasota–Yorke lemma (Theorem 1.2.4) and certain invariant measures that are constructed using Banach limits and have emerged in the work of Oxtoby and Ulam [54]; we call these measures *elementary*. Along with the KBBY decomposition we also extend Theorem 1 of Oxtoby and Ulam [54] to our setting.

In Section 2.2 we extend the notion of orbits used in dynamical systems to general Markov–Feller operators, and use these orbits to study the supports of the elementary invariant measures in Theorem 2.2.1, and (as mentioned before) to obtain "formulas" for the supports of the ergodic measures in Theorem 2.2.2 (as explained in the second paragraph after Corollary 2.2.3, the results of Theorem 2.2.1 and Theorem 2.2.2 where obtained by thinking of orbits as topological lower limits of suitable sequences of subsets of the locally compact separable metric space on which the Markov–Feller pairs are defined).

Finally, in Section 2.3 we specialize some of the results obtained in the first two sections of the chapter, in order to study certain Markov–Feller operators for which Skorokhod [64] uses the term *topologically connected,* and which we call *minimal* (the reason for our terminology is explained at the beginning of the section). In his paper [64] Skorokhod explains that the motivation for writing the work was the fact that minimality does not imply unique ergodicity even in the compact case; thus, a natural question is: can minimality be characterized by the supports of the invariant probabilities of Markov–Feller pairs (provided that the pairs under consideration have invariant probabilities, of course)? Our approach leads to an answer to the question, and to an extension of a known result in ergodic theory (see Theorem 6.17 of Walters [67]). We study the relationship between minimality and the property that each invariant probability be supported on the entire space.

All the results obtained in this chapter are used throughout the volume in the following chapters.

## 2.1   A Weak KBBY Decomposition

Throughout this section we assume given a Markov–Feller pair $(S, T)$ defined on a locally compact separable metric space $(X, d)$. We define a decomposition (which we call the weak KBBY decomposition) of $X$ generated by $(S, T)$, we study certain invariant measures that stem from the decomposition, and obtain necessary and sufficient conditions for the existence of $T$-invariant probabilities. The approach taken allows us to extend the classical KBBY decomposition to our setting.

Set

$$\Omega = \left\{ x \in X \ \middle| \ \begin{array}{c} L\left( (\langle f,\, T^n \delta_x \rangle)_{n \in \mathbb{N} \cup \{0\}} \right) = 0 \text{ for} \\ \text{every } f \in C_0(X) \text{ and every Banach limit } L \end{array} \right\}$$

and let $\Gamma = X \setminus \Omega$. We will refer to this splitting of $X$ into the sets $\Omega$ and $\Gamma$ as the *weak KBBY decomposition*, or the *$\Omega\Gamma$ decomposition*. Note that Theorem 1.3.2 implies that $\Gamma_0 \subseteq \Gamma$ where $\Gamma_0$ is the set defined in Section 1.2 ($\Gamma_0$ belongs to the classical KBBY decomposition).

For every $x \in \Gamma$ and every Banach limit $L$ let $\varepsilon_x^{(L)} : C_0(X) \to \mathbb{R}$ be defined by $\varepsilon_x^{(L)}(f) = L\left( (\langle f,\, T^n \delta_x \rangle)_{n \in \mathbb{N} \cup \{0\}} \right)$ for every $f \in C_0(X)$. It is easy to see that $\varepsilon_x^{(L)}$ is a positive linear functional, so $\varepsilon_x^{(L)}$ is also continuous. Accordingly, we may and do think of $\varepsilon_x^{(L)}$ as an element of $\mathcal{M}(X)$.

**Theorem 2.1.1.** *If $x \in \Gamma$ and $L$ is a Banach limit, then $\varepsilon_x^{(L)}$ is a $T$-invariant measure.*

*Proof.* Let $\phi : C_b(X) \to \mathbb{R}$ be defined by $\phi(f) = L\left( (\langle f,\, T^n \delta_x \rangle)_{n \in \mathbb{N} \cup \{0\}} \right)$ for every $f \in C_b(X)$. Clearly, $\phi$ is a positive linear functional, $\phi(1_X) = 1$, and the restriction of $\phi$ to $C_0(X)$ is $\varepsilon_x^{(L)}$. Since $L$ is a Banach limit, we obtain that

$$\phi(Sf) = L\left( (\langle Sf,\, T^n \delta_x \rangle)_{n \in \mathbb{N} \cup \{0\}} \right) = L\left( (\langle f,\, T^{n+1} \delta_x \rangle)_{n \in \mathbb{N} \cup \{0\}} \right)$$
$$= L\left( (\langle f,\, T^n \delta_x \rangle)_{n \in \mathbb{N} \cup \{0\}} \right) = \phi(f)$$

for every $f \in C_0(X)$ (actually, for every $f \in C_b(X)$). Thus, $\phi$ satisfies the conditions of the Lasota–Yorke lemma (Theorem 1.2.4); hence, we conclude that $T\varepsilon_x^{(L)} = \varepsilon_x^{(L)}$. $\square$

Given $x \in \Gamma$ and a Banach limit $L$, we call $\varepsilon_x^{(L)}$ an *elementary (T-)invariant measure (defined by $x$ and $L$)* whenever $\varepsilon_x^{(L)} \neq 0$.

Note that if $\varepsilon_x^{(L)}$ is an elementary invariant measure, then $0 < \|\varepsilon_x^{(L)}\| \leq 1$.

An elementary invariant measure may or may not be a probability. Indeed, if $x \in \Gamma_{cp}$, then the probability $\varepsilon_x$ defined in Section 1.2 is an elementary invariant measure; this is so because if $L'$ is a Banach limit, then $L : l^\infty \to \mathbb{R}$, $L\left((a_n)_{n \in \mathbb{N}}\right) = L'\left(\left(\frac{1}{n} \sum_{i=1}^{n} a_i\right)_{n \in \mathbb{N}}\right)$ for every $(a_n)_{n \in \mathbb{N}} \in l^\infty$ is also a Banach limit, and it is easy to see that $\varepsilon_x(f) = \varepsilon_x^{(L)}(f)$ for every $f \in C_0(X)$. By contrast, if $x \in \Gamma_c \setminus \Gamma_{cp}$, then $\varepsilon_x = \varepsilon_x^{(L)}$ (with $L$ as above), so $\varepsilon_x$ is an elementary invariant measure, but $\varepsilon_x$ is no longer a probability; a case in point appears in Example 1.2.8: the measure $\varepsilon_2$ is an elementary invariant measure (by the above discussion), but $\varepsilon_2$ is not a probability since $\varepsilon_2 = \frac{1}{2}\delta_1$.

The next theorem offers necessary and sufficient conditions for the existence of nonzero invariant measures for $(S, T)$.

**Theorem 2.1.2.** *The following assertions are equivalent:*

(a) *$T$ has invariant probabilities.*

(b) *$\Gamma_0 \neq \emptyset$.*

(c) *$\Gamma \neq \emptyset$.*

(d) *There exists $x_0 \in X$ and a compact subset $K$ of $X$ such that*

$$\limsup_{n \to \infty} \left(\frac{1}{n} \sum_{k=0}^{n-1} T^k \delta_{x_0}\right)(K) > 0.$$

*Proof.* (a) $\Rightarrow$ (b). Assume that $T$ has invariant probabilities and let $\mu^* \in \mathcal{M}(X)$ be such an invariant probability.

Since $\mu^*$ is a probability, there exists $f \in C_0(X)$ such that $f > 0$ and $\langle f, \mu^* \rangle > 0$. If $\Theta$ is the set defined in Corollary 1.2.7, then the corollary implies that $\Theta$ is nonempty. Thus, there exists $x_0 \in X$ such that the sequence $\left(\frac{1}{n} \sum_{k=0}^{n-1} S^k f(x_0)\right)_{n \in \mathbb{N}}$ converges to a (strictly) positive number; therefore, $\Gamma_0 \neq \emptyset$ since $x_0 \in \Gamma_0$.

(b) $\Rightarrow$ (c) is obvious since $\Gamma_0 \subseteq \Gamma$ by Theorem 1.3.2 as we pointed out earlier.

(c) $\Rightarrow$ (a) is also obvious since $\Gamma \neq \emptyset$ implies the existence of elementary $T$-invariant measures.

(a) $\Rightarrow$ (d) Let $\mu$ be a $T$-invariant probability (the existence of $\mu$ is assured by (a)). Then there exists $f \in C_0(X)$ such that $\langle f, \mu \rangle > 0$ and $0 \leq f(x) \leq 1$ for every $x \in X$; since $C_c(X)$ is dense in $C_0(X)$ we may and do pick $f \in C_c(X)$. By

Corollary 1.2.7 there exists $x_0 \in X$ such that the sequence $\left( \dfrac{1}{n} \sum\limits_{k=0}^{n-1} S^k f \left( x_0 \right) \right)_{n \in \mathbb{N}}$

converges, and $\lim\limits_{n \to \infty} \dfrac{1}{n} \sum\limits_{k=0}^{n-1} S^k f \left( x_0 \right) > 0$.

Let $K = \operatorname{supp} f$. Then

$$
0 < \lim_{n \to \infty} \left\langle \frac{1}{n} \sum_{k=0}^{n-1} S^k f, \delta_{x_0} \right\rangle = \lim_{n \to \infty} \left\langle f, \frac{1}{n} \sum_{k=0}^{n-1} T^k \delta_{x_0} \right\rangle
$$

$$
\leq \limsup_{n \to \infty} \left( \frac{1}{n} \sum_{k=0}^{n-1} T^k \delta_{x_0} \right) (K).
$$

(d) $\Rightarrow$ (b) Let $x_0 \in X$ and let $K$ be a compact subset of $X$ such that $\limsup\limits_{n \to \infty} \left( \dfrac{1}{n} \sum\limits_{k=0}^{n-1} T^k \delta_{x_0} \right) (K) > 0$.

By Proposition 7.1.8, p. 199 of Cohn's book [8] there exists $f \in C_c(X)$ such that $1_K \leq f \leq 1_X$.

We obtain that

$$
\limsup_{n \to \infty} \frac{1}{n} \sum_{k=0}^{n-1} S^k f \left( x_0 \right) = \limsup_{n \to \infty} \left\langle f, \frac{1}{n} \sum_{k=0}^{n-1} T^k \delta_{x_0} \right\rangle \geq \frac{1}{n} \sum_{k=0}^{n-1} T^k \delta_{x_0} (K) > 0.
$$

Thus, $x_0 \in \Gamma_0$.                                                                         $\square$

The equivalence of (a) and (b), and of (a) and (d) extend Theorem 1, p. 395 of Yosida's book [70], and Theorem 1 of Oxtoby and Ulam [54], respectively, to our setting.

Theorem 3.1 of Lasota and Yorke [42] implies that each of the four equivalent assertions of Theorem 2.1.2 is also equivalent to:

(e) *There exists a compact subset $K$ of $X$ and a probability $\mu_0 \in \mathcal{M}(X)$ such that* $\limsup\limits_{n \to \infty} \left( \dfrac{1}{n} \sum\limits_{k=0}^{n-1} T^k \mu_0(K) \right) > 0$.

(Note that even though Theorem 3.1 of [42] is proved under the assumption that the metric space $(X, d)$ has the property that every closed ball is a compact subset of $X$, the proof offered in [42] is valid in our setting, too (but using the version of Lemma 3.1 of [42] that we offer in Theorem 1.2.4, of course).)

For every $x \in X$ and $f \in C_0(X)$ set $f^*(x) = \lim\limits_{n \to \infty} \dfrac{1}{n} \sum\limits_{k=0}^{n-1} S^k f(x)$ whenever

the sequence $\left( \dfrac{1}{n} \sum\limits_{k=0}^{n-1} S^k f(x) \right)_{n \in \mathbb{N}}$     converges.

In view of the fact that $C_0(X)$ is separable (see Theorem 1.3.3) it is easy to see that the sets $\mathcal{D}$, $\Gamma_0$, $\Gamma_c$, and $\Gamma_{cp}$ belong to $\mathcal{B}(X)$.

Set

$$\Gamma_1 = \left\{ x \in \Gamma_{cp} \; \middle| \; \int_{\Gamma_{cp}} (f^*(y) - f^*(x))^2 \, d\varepsilon_x(y) = 0 \text{ for every } f \in C_0(X) \right\}.$$

Then $\Gamma_1 \in \mathcal{B}(X)$ (see Chap. 13, Section 4 of Yosida [70]). In general, $\Gamma_1 \neq \Gamma_{cp}$ (see Example 2.2.4).

We now define a relation $\sim$ on $\Gamma_1$ as follows: $x \sim y$ if and only if $f^*(x) = f^*(y)$ for every $f \in C_0(X)$. It is easy to see that $\sim$ is an equivalence relation on $\Gamma_1$. We will denote by $[x]$ the equivalence class of $x$ with respect to $\sim$ whenever $x \in \Gamma_1$.

The following theorem summarizes several results of Chap. 13, Section 4 of Yosida [70]:

**Theorem 2.1.3.** (a) *If $x \in \Gamma_1$, then $\varepsilon_x$ is an ergodic measure. Conversely, if $\mu \in \mathcal{M}(X)$ is an ergodic measure, then $\mu = \varepsilon_x$ for some $x \in \Gamma_1$.*

(b) *If $x \in \Gamma_1$, then the equivalence class $[x]$, as a subset of $X$, belongs to $\mathcal{B}(X)$, and $\varepsilon_x([x]) = 1$.*

If the space $(X, d)$ is compact, and the Markov–Feller pair $(S, T)$ is uniquely ergodic, then a well-known result (see, for example, Proposition 1.2, p. 178 of Krengel [32]) implies that $X = \Gamma_1 = [x]$ for every $x \in X$.

If $(X, d)$ is not compact and $(S, T)$ is uniquely ergodic, then $\Gamma_{cp} = \Gamma_1 = [x]$ for all $x \in \Gamma_1$; in this case $\Gamma_1$ may or may not be equal to $X$. For instance, if $X = \mathbb{N}$ and $(S, T)$ is the Markov–Feller pair of Example 1.1.14, then $(S, T)$ is uniquely ergodic, and the unique $T$-invariant probability is the Dirac measure concentrated at 1 (as pointed out in the subsection *Invariant Probabilities of Markov–Feller Operators* of Section 1.2); it is easy to see that $\Gamma_1 = \{2k - 1 | k \in \mathbb{N}\}$, so $\Gamma_1 \neq X$. The case $\Gamma_1 = X$ is illustrated in the following well-known example (as the reader will no doubt recognize, the example was used to obtain Example 1.1.14).

*Example 2.1.4.* Let $X = \mathbb{N}$, and let $d$ be as in Example 1.1.14. Let $T : l^1 \to l^1$ be defined as follows: if $\alpha \in l^1$, $\alpha = (x_n)_{n \in \mathbb{N}}$, then $T\alpha = (y_n)_{n \in \mathbb{N}}$ where

$$y_n = \begin{cases} x_1 + x_2 & \text{if} \quad n = 1 \\ x_{n+1} & \text{if} \quad n \geq 2 \end{cases}.$$

Thus, $T\left((x_n)_{n \in \mathbb{N}}\right) = (x_1 + x_2, \, x_3, \, x_4, \, x_5, \, \dots)$ for every $(x_n)_{n \in \mathbb{N}} \in l^1$.

If we let $S : l^\infty \to l^\infty$ be the adjoint of $T$, then $(S, T)$ is a Markov–Feller pair, and $S$ acts as follows: if $\gamma \in l^\infty$, $\gamma = (u_n)_{n \in \mathbb{N}}$, then $S\gamma = (v_n)_{n \in \mathbb{N}}$ where

$$v_n = \begin{cases} u_1 & \text{if} \quad n = 1 \text{ or } n = 2 \\ u_{n-1} & \text{if} \quad n \geq 3 \end{cases};$$

that is, $S\left((u_n)_{n\in\mathbb{N}}\right) = (u_1,\ u_1,\ u_2,\ u_3,\ u_4,\ \dots\ )$.

It is easy to see that $(S,T)$ is uniquely ergodic (the unique invariant probability of $(S,T)$ is the Dirac measure concentrated at 1), and that $\Gamma_1 = \mathbb{N} = [x]$ for every $x \in \mathbb{N}$.                                                                    ∎

The rotations of the unit circle (Example 1.1.11) can be used for a better understanding of the relationship between ergodic measures and the KBBY decomposition. Let $X = \mathbb{R}/\mathbb{Z}$, let $a \in \mathbb{R}/\mathbb{Z}$, and assume first that the equivalence class $a$ contains irrational numbers. Then $(S_a, T_a)$ is strictly ergodic, and the only invariant probability of $(S_a, T_a)$ is the Haar (Lebesgue) measure on $\mathbb{R}/\mathbb{Z}$ (as pointed out in the subsection *Invariant Probabilities of Markov–Feller Operators* of Section 1.2). Thus, in this case $X = \Gamma_{cp} = \Gamma_1 = [x]$ for every $x \in X$.

Assume now that the equivalence class $a \in \mathbb{R}/\mathbb{Z}$ contains rational numbers. If $a$ is the zero class (that is, if $a$ contains an integer), then $T_a$ is the identity operator, so $X = \Gamma_1$, and $[x] = \{x\}$ for every $x \in \Gamma_1$ (each equivalence class with respect to the equivalence relation $\sim$ on $\Gamma_1$ contains only one element). If $a$ is not the zero class, then there exists a rational number $\dfrac{p}{q}$ in the equivalence class $a$ such that $p \in \mathbb{N}$, $q \in \mathbb{N}$, $0 < \dfrac{p}{q} < 1$, and the greatest common divisor of $p$ and $q$ is 1. It is easy to see that in this case $\Gamma_1 = X$. By Theorem 2.1.3 the set of all ergodic measures is $\{\varepsilon_x \mid x \in X\}$. Each ergodic measure $\varepsilon_x$, $x \in X$ is defined by

$$\varepsilon_x\left(\left\{x + \frac{k}{q}\right\}\right) = \frac{1}{q} \ \text{for every } k = 0,\ 1,\ 2,\ \dots,\ q - 1.$$

**Extending the KBBY Decomposition to a More General Setting.** A natural research topic in connection with the KBBY decomposition is the study of the possibility of extending the decomposition to more general metric spaces $(X, d)$; that is, to try to relax the assumptions of local compactness and separability that we usually impose on $(X, d)$ in this book. An added incentive for such an investigation is the fact that since the mid-nineties there has been a constant interest in the study of Markov–Feller operators on Polish spaces (see, for example, the recent memoir of Szarek [66]). We will now outline briefly a possible approach to such an extension.

Thus, assume that $(X, d)$ is a metric space (not necessarily locally compact or separable), and let $\mathcal{B}(X)$, $\mathcal{M}(X)$, $C_b(X)$, and $B_b(X)$ be defined as in Section 1.1 (note that we can also define $C_0(X)$ as in Section 1.1, but here $C_0(X)$ does not play any significant role, except when $X$ is locally compact and separable, of course). Clearly, we can (and do) define the transition probabilities as in Section 1.1. If $P : X \times \mathcal{B}(X) \to \mathbb{R}$ is a transition probability, then it is easy to see that the map $T : \mathcal{M}(X) \to \mathcal{M}(X)$ defined by the equality (1.1.2) is well-defined, and is a Markov operator as defined in the subsection *Vector Lattices, Banach Lattices, and Positive Operators* of Section 1.3. Moreover, we can use the equality (1.1.3) in order to define an operator $S : B_b(X) \to B_b(X)$. Note that using arguments that appear in the proof of Proposition 1.1.4, we obtain that $S$ is well-defined (that is,

$Sf$ belongs to $B_b(X)$ whenever $f \in B_b(X)$). Clearly, $S$ is a positive contraction of $B_b(X)$, and $S$ and $T$ satisfy the equality (1.1.1) for every $f \in B_b(X)$ and $\mu \in \mathcal{M}(X)$. We say that $(S,T)$ is the *Markov pair defined by* $P$, and, in agreement with the terminology introduced in Section 1.1, we say that $(S,T)$ is the *Markov–Feller pair defined by* $P$ if, in addition, $S$ has the property that $Sf$ belongs to $C_b(X)$ whenever $f \in C_b(X)$.

Now let $P$ be a transition probability defined on $(X,d)$.

Assume that $(X,d)$ is a metric space that has the property that there exists an ideal $I$ of $C_b(X)$ such that:

(i) For every $x \in X$ there exists $u \in I$ such that $u(x) > 0$.

(ii) The norm of $C_b(X)$ induces on $I$ a Banach lattice structure (that is, when we consider on $I$ the pointwise order and the norm defined by the restriction of the uniform (sup) norm of $C_b(X)$ to $I$, then $I$ is a Banach lattice in its own right).

(iii) The topological dual $I'$ of $I$ is isometric and order isomorphic to $\mathcal{M}(X)$; therefore, we may and do identify the elements of $I'$ with the corresponding elements of $\mathcal{M}(X)$.

Then using Proposition 1.3.5 (the extension of Lemma 1.2.3 discussed in the subsection *Vector Lattices, Banach Lattices, and Positive Operators* of Section 1.3), we can extend Theorem 1.2.4 (the Lasota–Yorke lemma) to our setting here by replacing $C_0(X)$ by the ideal $I$.

Clearly, the sets $\mathcal{D}$, $\Gamma_c$, $\Omega$, $\Gamma$, $\Gamma_{cp}$, $\Gamma_1$, and the measures $\varepsilon_x^{(L)}$, where $L$ is a Banach limit and $x \in \Gamma$ can be defined with respect to the ideal $I$; for example, in our setting now

$$\Omega = \left\{ x \in X \;\middle|\; \begin{array}{l} L\left( (\langle f, T^n \delta_x \rangle)_{n \in \mathbb{N} \cup \{0\}} \right) = 0 \text{ for} \\ \text{every } f \in I \text{ and every Banach limit } L \end{array} \right\}.$$

It is easy to see (by using the above-mentioned extension of the Lasota–Yorke lemma, and by adapting the arguments of Theorem 2.1.1 to our current setting) that the measure $\varepsilon_x^{(L)}$ is $T$-invariant whenever $L$ is a Banach limit and $x \in \Gamma$.

If the ideal $I$ is also separable (that is, there exists a sequence $(f_l)_{l \in \mathbb{N}}$ of elements of $I$ such that the range $\{f_l \,|\, l \in \mathbb{N}\}$ of $(f_l)_{l \in \mathbb{N}}$ is dense in $I$ (i.e. $I \subseteq \overline{\{f_l \,|\, l \in \mathbb{N}\}}$)), then it can be shown that the sets $\mathcal{D}$, $\Gamma_0$, $\Gamma_c$, and $\Gamma_{cp}$ are measurable.

Our comments lead to the following natural open question: find a class $\mathcal{C}$ of metric spaces larger than the class of all locally compact separable metric spaces such that for every metric space in the class $\mathcal{C}$ there exists an ideal $I$ with the properties (i), (ii), and (iii) stated in this subsection.

## 2.2   Supports of Elementary Invariant and Ergodic Measures

In Section 2.1 we have defined the elementary invariant measures and have discussed the role played by the ergodic measures in the KBBY decomposition (because of this role, the decomposition is called an ergodic decomposition). Our goal now is to define the orbit of an element under the action of a Markov–Feller operator, and to use orbits in order to study the supports of measures that are elementary invariant or ergodic. In the case of ergodic measures, we will actually obtain "formulas" for the supports of such measures in terms of orbits.

As in the previous section, we assume given a Markov–Feller pair $(S, T)$ defined on a locally compact separable metric space $(X, d)$.

The *orbit of an element* $x \in X$ *under the action of* $T$ (or of $(S, T)$) is a subset of $X$ denoted $\mathcal{O}(x)$, and defined by $\mathcal{O}(X) = \bigcup_{n=0}^{\infty} \text{supp}\, (T^n \delta_x)$. The *orbit-closure of* $x$ (*under the action of* $T$ (or of $(S, T)$)) is the closure $\overline{\mathcal{O}(x)}$ of $\mathcal{O}(x)$ in the topology induced by the metric $d$.

If the Markov–Feller pair $(S, T)$ is induced by a continuous function $w : X \to X$ (see the discussion preceding Example 1.1.9), then $T^n \delta_x = \delta_{w^n(x)}$ for every $n \in \mathbb{N} \cup \{0\}$ and $x \in X$. Our terminology stems from the fact that in this case $\mathcal{O}(x)$ and $\overline{\mathcal{O}(x)}$, $x \in X$ are the usual forward orbits and forward orbit-closures that appear in the study of dynamical systems and in topological dynamics (see, for example, Furstenberg [20], Gottschalk and Hedlund [22], or Robinson [58]).

The next theorem deals with supports of $T$-invariant elementary measures of $(S, T)$ (whenever such measures exist, of course). Note that we do not impose the condition that $(S, T)$ be induced by a continuous function.

**Theorem 2.2.1.** *Assume that* $\Gamma \neq \emptyset$, *let* $x \in \Gamma$, *let* $L$ *be a Banach limit, and suppose that* $\varepsilon_x^{(L)}$ *is a* $T$-*invariant elementary measure. Then:*

(a) $supp\ \varepsilon_x^{(L)} \subseteq \overline{\mathcal{O}(x)}$.

(b) *If* $x \in supp\ \varepsilon_x^{(L)}$, *then* $supp\ \varepsilon_x^{(L)} = \overline{\mathcal{O}(x)}$.

*Proof.* (a) We have to prove that for every $z \in \text{supp}\ \varepsilon_x^{(L)}$ there exists a sequence $(y_k)_{k \in \mathbb{N}}$ of elements of $\mathcal{O}(x)$ such that $(y_k)_{k \in \mathbb{N}}$ converges to $z$ in the topology induced by the metric $d$ on $X$.

To this end, let $z \in \text{supp}\ \varepsilon_x^{(L)}$. Since $X$ is locally compact, there exists $\alpha \in \mathbb{R}$, $\alpha > 0$ such that $\overline{B(z, \alpha)}$ is compact in $X$.

Now let $(f_k)_{k \in \mathbb{N}}$ be a sequence of functions defined as follows: $f_k : X \to \mathbb{R}$, $f_k(y) = d\left(y, \left(X \setminus B\left(z, \frac{\alpha}{k}\right)\right)\right)$ for every $k \in \mathbb{N}$ and $y \in X$. Clearly, $f_k$ is continuous, and $\text{supp}\ f_k = \overline{B\left(z, \frac{\alpha}{k}\right)}$; therefore, $f_k \in C_c(X)$ for every $k \in \mathbb{N}$.

Since $z \in \operatorname{supp} \varepsilon_x^{(L)}$, it follows that $\varepsilon_x^{(L)}\left(B\left(z, \frac{\alpha}{k}\right)\right) > 0$; consequently, taking into consideration that $f_k(y) > 0$ if and only if $y \in B\left(z, \frac{\alpha}{k}\right)$, and that $f_k \in C_c(X) \subseteq C_0(X)$, we obtain that $0 < \left\langle f_k, \varepsilon_x^{(L)} \right\rangle = L\left( (\langle f_k, T^n \delta_x \rangle)_{n \in \mathbb{N} \cup \{0\}} \right)$ for every $k \in \mathbb{N}$.

Accordingly, for every $k \in \mathbb{N}$ there exists $n_k \in \mathbb{N} \cup \{0\}$ such that $\langle f_k, T^{n_k} \delta_x \rangle > 0$.

Since $f_k(y) > 0$ for every $y \in B\left(z, \frac{\alpha}{k}\right)$, it follows that $T^{n_k} \delta_x \left(B\left(z, \frac{\alpha}{k}\right)\right) > 0$; hence, $(\operatorname{supp}(T^{n_k}\delta_x)) \cap B\left(z, \frac{\alpha}{k}\right) \neq \emptyset$; therefore, there exists $y_k \in (\operatorname{supp}(T^{n_k}\delta_x)) \cap B\left(z, \frac{\alpha}{k}\right)$ for every $k \in \mathbb{N}$.

Clearly, the sequence $(y_k)_{k \in \mathbb{N}}$ converges to $z$, and $y_k \in \mathcal{O}(x)$ for every $k \in \mathbb{N}$.

(b) Clearly, in view of (a) we only have to prove that $\overline{\mathcal{O}(x)} \subseteq \operatorname{supp} \varepsilon_x^{(L)}$. Since $x \in \varepsilon_x^{(L)}$, Proposition 1.1.7 and the fact that $\varepsilon_x^{(L)}$ is a $T$-invariant measure imply that $(\operatorname{supp}(T^n \delta_x)) \subseteq \left(\operatorname{supp} \varepsilon_x^{(L)}\right)$ for every $n \in \mathbb{N} \cup \{0\}$; hence, $\mathcal{O}(x) \subseteq \operatorname{supp} \varepsilon_x^{(L)}$. Since the support of a measure is a closed set, it follows that $\overline{\mathcal{O}(x)} \subseteq \operatorname{supp} \varepsilon_x^{(L)}$. $\quad\square$

A natural question related to the statement of Theorem 2.2.1 is: does it occur that the inclusion at (a) of Theorem 2.2.1 is strict? (In other words, if $x \in \Gamma$ and the Banach limit $L$ are such that $\varepsilon_x^{(L)}$ is a $T$-invariant elementary measure, is it possible that $x$ does not belong to $\operatorname{supp} \varepsilon_x^{(L)}$?) If the answer were no, (that is, if $x \in \operatorname{supp} \varepsilon_x^{(L)}$ whenever $x$ and $L$ are such that $\varepsilon_x^{(L)}$ is a $T$-invariant elementary measure), then (a) of Theorem 2.2.1 would be redundant. However, the answer is yes; that is, it often occurs that the inclusion in Theorem 2.2.1-(a) is strict; Example 1.1.14, Example 1.2.8, and Example 2.1.4 can be used to illustrate this point. If $(S, T)$ is the Markov–Feller pair of Example 2.1.4 (therefore, $\Gamma = \mathbb{N}$ in this case), if $k \in \mathbb{N}$ is such that $k \geq 2$, and if $L$ is a Banach limit, then $\operatorname{supp} \varepsilon_k^{(L)} = \{1\}$ (because $\varepsilon_k^{(L)} = \delta_1$) while $\overline{\mathcal{O}(k)} = \mathcal{O}(k) = \{1, 2, 3, \ldots, k\}$. (Note that if $k = 1$, then $\overline{\mathcal{O}(1)} = \operatorname{supp} \varepsilon_1^{(L)} = \{1\}$, so we are in the case $(b)$ of Theorem 2.2.1.) Similarly, if $(S, T)$ is the Markov–Feller pair of Example 1.1.14 ($\Gamma = \{2k - 1 | k \in \mathbb{N}\}$ in this case), if $k \in \mathbb{N}$, $k \geq 2$ and $L$ is a Banach limit, then $\operatorname{supp} \varepsilon_{2k-1}^{(L)} = \{1\}$, while $\overline{\mathcal{O}(2k - 1)} = \mathcal{O}(2k - 1) = \{1, 3, 5, \ldots, 2k - 1\}$. Finally, let $(S, T)$ be the Markov–Feller pair of Example 1.2.8 (so, $\Gamma = \{2k - 1 | k \in \mathbb{N}\} \cup \{2\}$ in this case), let $k = 2$, and let $L$ be a Banach limit; then $\varepsilon_2^{(L)} = \frac{1}{2}\delta_1$, so $\operatorname{supp} \varepsilon_2^{(L)} = \{1\}$, while $\overline{\mathcal{O}(2)} = \mathcal{O}(2) = \{2k | k \in \mathbb{N}\} \cup \{1\}$. Note that in this case $\varepsilon_2^{(L)}$ is a $T$-invariant elementary measure, but not a probability. Note also that Example 1.2.8 can be used to illustrate that, in general, $\Gamma$ is not orbit-invariant (that is, in general, $x \in \Gamma$ does not imply that $\mathcal{O}(x) \subseteq \Gamma$) since $2 \in \Gamma$ but $\mathcal{O}(2) \not\subseteq \Gamma$ in the example.

We now turn our attention to ergodic measures. The next theorem offers a "formula" for the support of an ergodic measure. In the theorem we use the notations introduced before Theorem 2.1.3.

**Theorem 2.2.2.** *Assume that* $\Gamma_1 \neq \emptyset$, *and let* $x \in \Gamma_1$. *Then* $\operatorname{supp} \varepsilon_x = \bigcap_{y \in [x]} \overline{\mathcal{O}(y)}$.

*Proof.* We have to prove that:

(a) $\operatorname{supp} \varepsilon_x \subseteq \overline{\mathcal{O}(y)}$ for every $y \in [x]$        and

(b) $\operatorname{supp} \varepsilon_x \supseteq \bigcap_{y \in [x]} \overline{\mathcal{O}(y)}$.

(a) Let $y \in [x]$ and let $z \in \operatorname{supp} \mu$.
Since we have to prove that $z \in \overline{\mathcal{O}(y)}$, we will construct a sequence $(z_n)_{n \in \mathbb{N}}$ of elements of $\mathcal{O}(y)$ such that $(z_n)_{n \in \mathbb{N}}$ converges to $z$. The construction of $(z_n)_{n \in \mathbb{N}}$ is to a large extent similar to the construction of the sequence $(y_n)_{n \in \mathbb{N}}$ in the proof of (a) of Theorem 2.2.1. Thus, let $\alpha \in \mathbb{R}$, $\alpha > 0$, be such that $\overline{B(z, \alpha)}$ is compact, and let $(f_n)_{n \in \mathbb{N}}$ be a sequence of functions defined as follows: $f_n : X \to \mathbb{R}$, $f_n(t) = d\left(t, X \setminus B\left(z, \frac{\alpha}{n}\right)\right)$ for every $n \in \mathbb{N}$ and $t \in X$.

As in the proof of Theorem 2.2.1-$(a)$, it follows that $f_n \in C_c(X)$ and $\operatorname{supp} f_n = \overline{B\left(z, \frac{\alpha}{n}\right)}$ for every $n \in \mathbb{N}$. Taking into consideration that $\mu\left(\overline{B\left(z, \frac{\alpha}{n}\right)}\right) > 0$ (since $z \in \operatorname{supp} \mu$), we obtain that $\langle f_n, \mu \rangle > 0$ for every $n \in \mathbb{N}$.

Since $y \in [x]$, it follows that the sequence $\left(\frac{1}{l} \sum_{k=0}^{l-1} S^k f_n(y)\right)_{l \in \mathbb{N}}$ converges to $\langle f_n, \mu \rangle$; consequently,

$$0 < \langle f_n, \mu \rangle = \lim_{l \to \infty} \left\langle \frac{1}{l} \sum_{k=0}^{l-1} S^k f_n, \delta_y \right\rangle = \lim_{l \to \infty} \left\langle f_n, \frac{1}{l} \sum_{k=0}^{l-1} T^k \delta_y \right\rangle$$

for every $n \in \mathbb{N}$.

Thus, given $n \in \mathbb{N}$ there exists $l_n \in \mathbb{N}$ such that $\left\langle f_n, \frac{1}{l_n} \sum_{k=0}^{l_n-1} T^k \delta_y \right\rangle > 0$.

Since $(\operatorname{supp} f_n) = \overline{B\left(z, \frac{\alpha}{n}\right)}$, it follows that $\frac{1}{l_n} \sum_{k=0}^{l_n-1} T^k \delta_y \left(\overline{B\left(z, \frac{\alpha}{n}\right)}\right) > 0$; hence,

$$\overline{B\left(z, \frac{\alpha}{n}\right)} \cap \left(\bigcap_{k=0}^{l_n-1} \left(\operatorname{supp} \left(T^k \delta_y\right)\right)\right) \neq \emptyset.$$

Let $z_n \in \overline{B\left(z, \frac{\alpha}{n}\right)} \cap \left(\bigcap_{k=0}^{l_n-1} \left(\operatorname{supp} \left(T^k \delta_y\right)\right)\right)$.

Clearly, $z_n \in \mathcal{O}(y)$ for every $n \in \mathbb{N}$, and the sequence $(z_n)_{n \in \mathbb{N}}$ converges to $z$.

Thus, $(\operatorname{supp} \mu) \subseteq \overline{\mathcal{O}(y)}$ for every $y \in [x]$.

(b) Since $\varepsilon_x([x]) = 1$ (by Theorem 2.1.3-(b)), and since $\varepsilon_x(\operatorname{supp} \varepsilon_x) = 1$, it follows that $\varepsilon_x([x] \cap (\operatorname{supp} \varepsilon_x)) = 1$, so there exists $z \in [x] \cap (\operatorname{supp} \varepsilon_x)$. Using Proposition 1.1.7, we obtain that $(\operatorname{supp} T^n \delta_z) \subseteq (\operatorname{supp} T^n \varepsilon_x) = (\operatorname{supp} \varepsilon_x)$ for every $n \in \mathbb{N} \cup \{0\}$; therefore, $\mathcal{O}(z) \subseteq (\operatorname{supp} \varepsilon_x)$. Accordingly, $\bigcap_{y \in [x]} \mathcal{O}(y) \subseteq$ $(\operatorname{supp} \varepsilon_x)$. $\qquad\square$

The proof of (b) of the above theorem has the following consequence:

**Corollary 2.2.3.** *If $\Gamma_1 \neq \emptyset$, and if $x \in \Gamma_1$, then $\operatorname{supp} \varepsilon_x = \overline{\mathcal{O}(y)}$ whenever $y \in$ $[x] \cap (\operatorname{supp} \varepsilon_x)$.*

Note that both Theorem 2.2.2 and Corollary 2.2.3 offer "formulas" for the support of any ergodic measure since by Theorem 2.1.3-(a) any ergodic measure is of the form $\varepsilon_x$ for some $x \in \Gamma_1$.

Any orbit-closure is the topological lower limit of a suitable sequence of subsets of $X$; to be precise, $\overline{\mathcal{O}(x)} = \underset{n \to \infty}{\operatorname{Li}} \left( \bigcup_{k=0}^{n} \left( \operatorname{supp} \left( T^k \delta_x \right) \right) \right)$ whenever $x \in X$. Theorem 2.2.1, Theorem 2.2.2, and Corollary 2.2.3 were obtained by thinking of orbits in terms of topological lower limits.

We saw in Theorem 2.2.1 that the elementary invariant probability measures $\varepsilon_x$, $x \in \Gamma_{cp}$ are "minimal" in the sense that given $x \in \Gamma_{cp}$, then $\operatorname{supp} \varepsilon_x$ is always included in the orbit-closure $\overline{\mathcal{O}(x)}$ of $x$, and if $x \in \operatorname{supp} \varepsilon_x$, then $\operatorname{supp} \varepsilon_x = \overline{\mathcal{O}(x)}$. On the other hand, an ergodic measure $\mu$ is "minimal," as well, in the sense that if $A_1$ and $A_2$ are two measurable subsets of $\operatorname{supp} \mu$ such that $A_1 \cup A_2 = \operatorname{supp} \mu$, $A_1 \cap A_2 = \emptyset$, $\mu(A_1) > 0$, and $\mu(A_2) > 0$, then the measures $\nu_1$ and $\nu_2$ defined by $\nu_i(A) = \mu(A_i \cap A)$ for every $A \in \mathcal{B}(X)$ and $i = 1, 2$ cannot be $T$-invariant measures. Thus, we may be tempted to believe that $\Gamma_{cp} = \Gamma_1$. The following surprisingly simple example suggested by one of the anonymous referees of this book shows that, in general, $\Gamma_1$ is a proper subset of $\Gamma_{cp}$.

*Example 2.2.4.* Let $X = \{1, 2, 3\}$, and let $d$ be the distance on $X$ defined by $d(i, j) = |i - j|$ for every $i \in \{1, 2, 3\}$ and $j \in \{1, 2, 3\}$. Then $(X, d)$ is a separable compact metric space, and $C_0(X) = C_b(X) = B_b(X)$.

It is easy to see that each of the spaces $C_b(X)$ and $\mathcal{M}(X)$ is isometric and order isomorphic to $\mathbb{R}^3$, so we think of the elements of $C_b(X)$ as row vectors of $\mathbb{R}^3$, and of the elements of $\mathcal{M}(X)$ as column vectors of $\mathbb{R}^3$ as follows: to each $f \in C_b(X)$ we associate the vector $(f(1), f(2), f(3))$ in $\mathbb{R}^3$, and conversely, for each vector $(u_1, u_2, u_3) \in \mathbb{R}^3$ we define $f \in C_b(X)$ as follows: $f(1) = u_1$, $f(2) = u_2$, and $f(3) = u_3$; similarly, to each $\mu \in \mathcal{M}(X)$ we associate the vector $\begin{pmatrix} \mu(\{1\}) \\ \mu(\{2\}) \\ \mu(\{3\}) \end{pmatrix}$ in $\mathbb{R}^3$, and conversely, for every $\begin{pmatrix} x \\ y \\ z \end{pmatrix}$ in $\mathbb{R}^3$ we define $\mu \in \mathcal{M}(X)$ by $\mu(\{1\}) = x$, $\mu(\{2\}) = y$, and $\mu(\{3\}) = z$.

Now consider the matrix $A = \begin{pmatrix} 0 & 0 & 0 \\ \dfrac{1}{2} & 1 & 0 \\ \dfrac{1}{2} & 0 & 1 \end{pmatrix}$ and define two operators $S$ and $T$ as

follows: $S : C_b(X) \to C_b(X)$, $Sf$ corresponds to the row vector $(f(1), f(2), f(3)) \cdot A$ for every $f \in C_b(X)$; $T : \mathcal{M}(X) \to \mathcal{M}(X)$, $T\mu$ corresponds to the column vector

$A \cdot \begin{pmatrix} \mu(\{1\}) \\ \mu(\{2\}) \\ \mu(\{3\}) \end{pmatrix}$ for every $\mu \in \mathcal{M}(X)$.

It is easy to see that $(S, T)$ is a Markov–Feller pair defined on $(X, d)$.

Now note that since $T\delta_1$ corresponds to $A \cdot \begin{pmatrix} 1 \\ 0 \\ 0 \end{pmatrix} = \begin{pmatrix} 0 \\ \dfrac{1}{2} \\ \dfrac{1}{2} \end{pmatrix}$, and since

$A \cdot \begin{pmatrix} 0 \\ \dfrac{1}{2} \\ \dfrac{1}{2} \end{pmatrix} = \begin{pmatrix} 0 \\ \dfrac{1}{2} \\ \dfrac{1}{2} \end{pmatrix}$ we obtain that $1 \in \Gamma_{cp}$ and $\varepsilon_1 = \dfrac{1}{2}\delta_2 + \dfrac{1}{2}\delta_3$. Since $T\delta_i = \delta_i$

for every $i = 2, 3$, it follows that $\varepsilon_i = \delta_i$, $i = 2, 3$, and that $\varepsilon_2$ and $\varepsilon_3$ are ergodic measures. Since $\varepsilon_1$ is not an ergodic measure, we obtain that $\Gamma_1 = \{2, 3\}$, while $\Gamma_{cp} = X$ in this case. ∎

We will conclude this section by discussing symbolic flows; that is, we will illustrate some of the results obtained so far in the case in which the Markov–Feller pair under consideration is induced by a symbolic flow (see Example 1.1.13).

To this end, let $l \in \mathbb{N}$, $l \geq 2$, and let $\Lambda = \{0, 1, \ldots, l - 1\}$. Also, let $(X, d)$ be the metric space defined in Example 1.1.13; that is, $X = \Lambda^{\mathbb{N}} =$ the set of all sequences of elements of $\Lambda$, and $d$ is the metric defined by

$$d\left((i_k)_{k \in \mathbb{N}}, (j_k)_{k \in \mathbb{N}}\right) = \sum_{k=1}^{\infty} 2^{-k}|i_k - j_k|$$

for every $(i_k)_{k \in \mathbb{N}} \in X$ and $(j_k)_{k \in \mathbb{N}} \in X$. The set $\Lambda$ is often called an *alphabet*; the elements of $\Lambda$ are called *letters*.

Let $w : X \to X$ be the map defined in Example 1.1.13, and let $(S_X, T_X)$ be the Markov–Feller pair defined by $w$ on $(X, d)$.

Following the well established terminology in the area (see, for example, Furstenberg [20]), if $\mathbf{w} \in \Lambda^m$ for some $m \in \mathbb{N}$, we call $\mathbf{w}$ a *word (of length $m$ with letters in $\Lambda$)*.

If $\omega \in \Lambda^{\mathbb{N}}$, $\omega = (i_k)_{k\in\mathbb{N}}$, and $\mathbf{w}$ is a word of length $m$, $m \in \mathbb{N}$, $\mathbf{w} = (j_1, j_2, \ldots, j_m)$, we say that $\omega$ *starts (or begins) with (the word)* $\mathbf{w}$ if $i_1 = j_1, i_2 = j_2, \ldots, i_m = j_m$.

Let $\mathbf{w}$ be a word. The function $f_{\mathbf{w}}(\omega) : \Lambda^{\mathbb{N}} \to \mathbb{R}$ defined by

$$f_{\mathbf{w}}(\omega) = \begin{cases} 1 \text{ if } \omega \text{ begins with } \mathbf{w} \\ 0 \text{ otherwise} \end{cases}$$

is called the *characteristic function of the word* $\mathbf{w}$. Note that for every word $\mathbf{w}$ the function $f_{\mathbf{w}}$ is continuous.

**Lemma 2.2.5.** *The set of all linear combinations of functions of the form $f_{\mathbf{w}}$, $\mathbf{w} \in \Lambda^m$, $m \in \mathbb{N}$ is dense in $C_b(X)$.*

*Proof.* Let $f \in C_b(X)$ and let $\varepsilon \in \mathbb{R}$, $\varepsilon > 0$. Since $X$ is compact, $f$ is uniformly continuous. Thus, there exists $\delta \in \mathbb{R}$, $\delta > 0$, such that $|f(\omega) - f(\omega')| < \dfrac{\varepsilon}{2}$ whenever $d(\omega, \omega') < \delta$. In view of the way in which $d$ is defined, it follows that there exists $m \in \mathbb{N}$ such that $d(\omega, \omega') < \delta$ whenever $\omega$ and $\omega'$ agree on their first $m$ components (that is, whenever there exists a word $\mathbf{w}$ of length $m$ such that both $\omega$ and $\omega'$ begin with $\mathbf{w}$).

For every word $\mathbf{w}$ of length $m$, let $\omega_{\mathbf{w}}$ be an element of $X$ that starts with $\mathbf{w}$, and set $g = \sum\limits_{\mathbf{w}\in\Lambda^m} f(\omega_{\mathbf{w}}) f_{\mathbf{w}}$.

Clearly, $g$ is a linear combination of the functions $f_{\mathbf{w}}$, $\mathbf{w} \in \Lambda^m$, and it is easy to see that $\|f - g\| \leq \dfrac{\varepsilon}{2} < \varepsilon$.

Thus, the set of all linear combinations of functions of the form $f_{\mathbf{w}}$, $\mathbf{w} \in \Lambda^m$, $m \in \mathbb{N}$ is dense in $C_b(X)$. $\qquad\square$

If $A$ is a finite set, we denote by $\operatorname{card}(A)$ the number of elements in $A$ (if $A$ is empty, $\operatorname{card}(A) = 0$, of course).

Let $\omega \in X$, let $m \in \mathbb{N}$, and let $\mathbf{w} \in \Lambda^m$. We say that $\mathbf{w}$ is a *regular word for $\omega$* if the sequence $\left( \dfrac{\operatorname{card}(\{k \in \mathbb{Z} \,|\, 0 \leq k \leq n, \ w^k(\omega) \text{ begins with } \mathbf{w}\})}{n} \right)_{n\in\mathbb{N}}$ converges.

We say that $\mathbf{w}$ is a *strongly regular word for $\omega$* if $\mathbf{w}$ is a regular word (for $\omega$), and

$$\lim_{n\to\infty} \frac{\operatorname{card}(\{k \in \mathbb{Z} \,|\, 0 \leq k \leq n, \ w^k(\omega) \text{ begins with } \mathbf{w}\})}{n} > 0.$$

The next theorem offers a characterization of the elements of the set $\Gamma_{cp}$ that appears in the KBBY decomposition of $(S_X, T_X)$. By the same token, since $\Gamma_0 = \Gamma = X$ (because $X$ is compact), the theorem allows us to characterize the elements of $\Gamma_0$ (or $\Gamma$) that belong to $\Gamma_{cp}$ in this case.

**Theorem 2.2.6.** *Let $\omega \in X$. Then $\omega$ belongs to $\Gamma_{cp}$ if and only if every word with letters in $\Lambda$ is a regular word for $\omega$.*

*Proof.* Assume first that $\omega$ has the property that every word with letters in $\Lambda$ is a regular word for $\omega$. In view of Lemma 2.2.5, we obtain that, in order to prove that

$\omega \in \Gamma_{cp}$ it is enough to prove that the sequence $\left( \dfrac{1}{n} \displaystyle\sum_{k=0}^{n-1} S^k f_{\mathbf{w}}(\omega) \right)_{n \in \mathbb{N}}$ converges whenever $\mathbf{w}$ is a word with letters in $\Lambda$.

Now, given such a word $\mathbf{w}$, note that

$$\frac{1}{n} \sum_{k=0}^{n-1} S^k f_{\mathbf{w}}(\omega) = \frac{\operatorname{card}\{k \in \mathbb{Z} \mid 0 \le k \le n, \, w^k(\omega) \text{ begins with } \mathbf{w}\}}{n} \qquad (2.2.1)$$

for every $n \in \mathbb{N}$; hence, the sequence $\left( \dfrac{1}{n} \displaystyle\sum_{k=0}^{n-1} S^k f_{\mathbf{w}}(\omega) \right)_{n \in \mathbb{N}}$ converges.

Conversely, if $\omega \in \Gamma_{cp}$, then the sequence $\left( \dfrac{1}{n} \displaystyle\sum_{k=0}^{n-1} S^k f_{\mathbf{w}}(\omega) \right)_{n \in \mathbb{N}}$ converges for every word $\mathbf{w}$; therefore, using (2.2.1) it is easy to see that every word is regular. $\qquad \square$

We know from Theorem 2.2.1 that if $\omega \in \Gamma_{cp}$ (that is, in view of Theorem 2.2.6, if every word is regular in $\omega$) that $\operatorname{supp} \varepsilon_\omega \subseteq \overline{\mathcal{O}(\omega)}$, and that $\operatorname{supp} \varepsilon_\omega = \overline{\mathcal{O}(\omega)}$ whenever $\omega \in \operatorname{supp} \varepsilon_\omega$. Thus, it is of interest to know under what conditions $\omega$ belongs to $\operatorname{supp} \varepsilon_\omega$. The next theorem offers such a necessary and sufficient condition.

**Theorem 2.2.7.** *Let* $\omega \in \Gamma_{cp}$, $\omega = (i_k)_{k \in \mathbb{N}}$. *Then* $\omega \in \operatorname{supp} \varepsilon_\omega$ *if and only if for every* $m \in \mathbb{N}$ *the word* $\mathbf{w}_m = (i_1, i_2, \ldots, i_m)$ *(the word formed of the first $m$ letters of $\omega$) is a strongly regular word for* $\omega$.

*Proof.* For every word $\mathbf{w}$, let $R_{\mathbf{w}}$ be the set of all $\omega \in X$ that start with $\mathbf{w}$. Then the collection of all $R_{\mathbf{w}}$ where $\mathbf{w}$ is a word is a basis for the topology defined by the metric $d$ on $X$.

Now assume that $\omega \in \Gamma_{cp}$, $\omega = (i_k)_{k \in \mathbb{N}}$ is such that $\omega \in \operatorname{supp} \varepsilon_\omega$. Then $R_{\mathbf{w}_m} \cap \operatorname{supp} \varepsilon_\omega \ne \emptyset$ for every $m \in \mathbb{N}$ where $\mathbf{w}_m = (i_1, i_2, \ldots, i_m)$. Thus, $\varepsilon_\omega(R_{\mathbf{w}_m}) > 0$; that is, $\mathbf{w}_m$ is a strongly regular word for $\omega$ for every $m \in \mathbb{N}$.

Conversely, assume that $\omega \in \Gamma_{cp}$, $\omega = (i_k)_{k \in \mathbb{N}}$ is such that $\mathbf{w}_m = (i_1, i_2, \ldots, i_m)$ is strongly regular for $\omega$ for every $m \in \mathbb{N}$.

Then, $\varepsilon_\omega(R_{\mathbf{w}_m}) > 0$ for every $m \in \mathbb{N}$. Since the collection $\{R_{\mathbf{w}} \mid \mathbf{w} \text{ is a word}\}$ is a basis for the topology of $X$, it follows that for every open set $G$ in $X$ such that $\omega \in G$ there exists $m \in \mathbb{N}$ such that $\omega \in R_{\mathbf{w}_m} \subseteq G$, so $\varepsilon_\omega(G) > 0$. Therefore, $\omega \in \operatorname{supp} \varepsilon_\omega$. $\qquad \square$

## 2.3   Minimal Markov–Feller Pairs

Our goal in this section is to use the results obtained so far in this chapter in order to study certain Markov–Feller pairs which we call minimal (a term borrowed from dynamical systems and topological dynamics), or topologically connected (the term used by Skorokhod [64]).

As always in this chapter, throughout this section we assume given a Markov–Feller pair $(S, T)$ defined on a locally compact separable metric space $(X, d)$.

We say that $T$ (or $(S, T)$) is *minimal* or *topologically connected* if every orbit is dense in $X$ (that is, if $\mathcal{O}(x)$ is dense in $X$ whenever $x \in X$).

We use the term minimal because if $(S, T)$ is induced by a continuous function, say, $w : X \rightarrow X$, then our comments on orbits made at the beginning of Section 2.2 imply that $(S, T)$ is minimal if and only if the set $\{w^n(x) | n \in \mathbb{N} \cup \{0\}\}$ is dense in $X$ for every $x \in X$ if and only if $X$ is a minimal set in the sense used in dynamical systems and topological dynamics (see, for example, p. 24 of Robinson's book [58]).

For the next proposition, let $P_n$ be the transition probability that generates the Markov–Feller pair $(S^n, T^n)$, $n \in \mathbb{N}$.

**Proposition 2.3.1.** *The following assertions are equivalent:*

(a) *For every* $x \in X$ *the orbit* $\mathcal{O}(x)$ *is dense in* $X$.

(b) $\sum\limits_{n=1}^{\infty} P^n(x, U) > 0$ *for every* $x \in X$ *and every open nonempty subset* $U$ *of* $X$.

*Proof.* (a)$\Rightarrow$(b) Let $x \in X$ and let $U$ be an open nonempty subset of $X$. Since $\mathcal{O}(x)$ is dense in $X$, it follows that $\left( \bigcup\limits_{n=0}^{\infty} \text{supp}\, (T^n \delta_x) \right) \cap U \neq \emptyset$.

In order to complete the proof of the implication, it is enough to prove that

$$(\text{supp}\, (T^n \delta_x)) \cap U \neq \emptyset \tag{2.3.1}$$

for some $n \in \mathbb{N}$. Indeed, if $n \in \mathbb{N}$ is such that (2.3.1) holds true, then $T^n \delta_x(U) > 0$; using (1.1.2) we obtain that $T^n \delta_x(U) = P_n(x, U)$; it is obvious now that we only have to prove that (2.3.1) holds true for some $n \in \mathbb{N}$.

Assume that $(\text{supp}\, (T^n \delta_x)) \cap U = \emptyset$ for every $n \in \mathbb{N}$. Then $\mathcal{O}(x) \cap U = \{x\}$.

It follows that $U = \{x\}$. Indeed, if there exists $z \in U$, $z \neq x$, then there exists $r \in \mathbb{R}$, $r > 0$ such that $x \notin B(z, r)$, and $B(z, r) \subset U$. But then $(\text{supp}\, (T^n \delta_x)) \cap U \supseteq \mathcal{O}(x) \cap B(z, r) \neq \emptyset$, so we obtain a contradiction since we have assumed that (2.3.1) is false for every $n \in \mathbb{N}$.

Now let $y \in \text{supp}\, T\delta_x$. Since we assume that (a) is true, it follows that $x \in \overline{\mathcal{O}(y)}$, so there exists a sequence $(y_k)_{k \in \mathbb{N}}$ of elements of $\mathcal{O}(y)$ such that $(y_k)_{k \in \mathbb{N}}$ converges to $x$. Since we assume that (2.3.1) is false for every $n \in \mathbb{N}$, it follows that $(\text{supp}\, (T\delta_x)) \cap U = \emptyset$, so $y \neq x$; hence, we may and do assume that $y_k \neq y$ for every $k \in \mathbb{N}$. Since $\{x\}$ is open, it follows that $y_{k_0} = x$ for some $k_0 \in \mathbb{N}$.

Clearly, $x = y_{k_0} \in \text{supp}\, (T^{n_{k_0}} \delta_y)$ for some $n_{k_0} \in \mathbb{N}$ because $y_{k_0} \in \mathcal{O}(y)$ and $y_{k_0} \neq y$.

Since $y \in \text{supp}\, \delta_y \subseteq \text{supp}\, (T\delta_x)$, Proposition 1.1.7 implies that $x \in \text{supp}\, (T^{n_{k_0}} \delta_y) \subseteq \text{supp}\, (T^{n_{k_0}+1} \delta_x)$.

We have obtained a contradiction since $x \in U$ and we have assumed that (2.3.1) is false for every $n \in \mathbb{N}$.

(b)$\Rightarrow$(a) Assume that $x \in X$ and that $\mathcal{O}(x)$ is not dense in $X$. Then $U = X \setminus \overline{\mathcal{O}(x)}$ is a nonempty open subset of $X$. Using (b) we obtain that $\sum_{n=1}^{\infty} P_n(x, U) > 0$; that is, $P_m(x, U) > 0$ for some $m \in \mathbb{N}$. Since $P_m(x, U) = T^m \delta_x(U)$, it follows that $U \cap \operatorname{supp}(T^m \delta_x) \neq \emptyset$. We obtained a contradiction which stems from the assumption that $\overline{\mathcal{O}(x)} \neq X$. $\quad\square$

Condition (b) of Proposition 2.3.1 appears in the paper [64] by Skorokhod, who uses the term topologically connected for the condition being satisfied.

It is tempting to believe (see Skorokhod [64]) that if the Markov–Feller pair $(S, T)$ is minimal, and if $(S, T)$ has nonzero finite invariant measures (for example, if the metric space $(X, d)$ is compact), then $(S, T)$ is uniquely ergodic. However, this is not true; that is, there exist minimal Markov–Feller pairs defined on compact metric spaces that are not uniquely ergodic (see Keane [30] and Boshernitzan [6], or Skorokhod [64]). Intuition might also suggest that if $(S, T)$ is minimal and has invariant probabilities, then each such invariant probability is supported on the entire space; as the next proposition shows, this "feeling" is correct.

**Proposition 2.3.2.** *Assume that the Markov–Feller pair $(S, T)$ is minimal and has invariant probabilities. If $\mu$ is an invariant probability for $(S, T)$, then $\operatorname{supp} \mu = X$.*

*Proof.* If $x \in \operatorname{supp} \mu$, then $\operatorname{supp}(T^n \delta_x) \subseteq \operatorname{supp}(T^n \mu) = \operatorname{supp} \mu$ by Proposition 1.1.7. Consequently, $\mathcal{O}(x) \subseteq \operatorname{supp} \mu$. Since $\overline{\mathcal{O}(x)} = X$, it is obvious that $\operatorname{supp} \mu = X$. $\quad\square$

Under a mild additional condition on $(S, T)$ we get the following converse of Proposition 2.3.2:

**Proposition 2.3.3.** *Assume that $X = \Gamma$ and that $\operatorname{supp} \mu = X$ whenever $\mu \in \mathcal{M}(X)$, $\mu$ is a probability, and $T\mu = \mu$. Then $(S, T)$ is a minimal Markov–Feller pair.*

*Proof.* Let $x \in X$. We have to prove that $\overline{\mathcal{O}(x)} = X$.

Since $x \in \Gamma$ (because $X = \Gamma$) there exists a Banach limit $L$ such that the measure $\varepsilon_x^{(L)}$ defined at the beginning of Section 2.1 (before Theorem 2.1.1) is a $T$-invariant elementary measure. Since $T\varepsilon_x^{(L)} = \varepsilon_x^{(L)}$, our assumption concerning the support of a $T$-invariant probability implies that $\operatorname{supp} \varepsilon_x^{(L)} = X$. Using Theorem 2.2.1 we obtain that $\overline{\mathcal{O}(x)} = X$. $\quad\square$

It is convenient to combine Proposition 2.3.2 and Proposition 2.3.3 into a single result as follows:

**Theorem 2.3.4.** *Assume that $X = \Gamma$. Then the following assertions are equivalent:*

(a) *The Markov–Feller pair $(S, T)$ is minimal.*

(b) *$\operatorname{supp} \mu = X$ whenever $\mu \in \mathcal{M}(X)$ and $\mu$ is a $T$-invariant probability.*

If $(X, d)$ is compact, then $\Gamma_0 = X$ (as pointed out in the subsection *The KBBY Decomposition* of Section 1.2). Since $\Gamma_0 \subseteq \Gamma$ (see the beginning of Section 2.1), it follows that $\Gamma = X$. Thus, in the compact case Theorem 2.3.4 becomes:

**Corollary 2.3.5.** *If $(X, d)$ is a compact metric space, then the following assertions are equivalent:*

(a) $(S, T)$ *is minimal.*

(b) *If $\mu \in \mathcal{M}(X)$ is a $T$-invariant probability, then supp $\mu = X$.*

Corollary 2.3.5 complements results of Skorokhod [64], and extends a known theorem in ergodic theory (see Theorem 6.17 of the book by Walters [67]) which states that (a) and (b) of the corollary are equivalent whenever $(S, T)$ is induced by a homeomorphism.

If $(S, T)$ is a Markov–Feller pair defined on a compact metric space $(X, d)$, then $(S, T)$ is minimal and uniquely ergodic if and only if $(S, T)$ is strictly ergodic. An example is $(S_a, T_a)$ of Example 1.1.11 whenever the equivalence class $a \in \mathbb{R}/\mathbb{Z}$ contains irrational numbers.

The minimal symbolic flows mentioned at Example 1.1.13 are sometimes uniquely ergodic (that is, strictly ergodic), and sometimes they are not. We mentioned earlier the works of Keynes and Newton [31] and of Keane [30]; they construct examples of interval exchange maps which are minimal, and each map has two distinct ergodic measures. The interval exchange maps do not generate a Markov–Feller pair induced by a continuous function as described before Example 1.1.9 simply because the interval exchange maps are not continuous. However, as pointed out by Boshernitzan on p. 78 of [6] any minimal interval exchange map of a finite number of intervals has an isomorphic representation as a minimal symbolic flow, so the examples of Keynes and Newton [31] and of Keane [30] yield examples of minimal symbolic flows that are not uniquely ergodic.

We say that the Markov–Feller pair $(S, T)$ is *trivially minimal* if supp $T\delta_x = X$ for every $x \in X$. One could ask whether or not there exist trivially minimal Markov–Feller pairs. Even though none of the examples of Markov–Feller pairs discussed so far is trivially minimal, common sense tells us that such Markov–Feller pairs do exist. Our goal now is to discuss such examples, but in order to do so we need some preparation.

Let $\mu_0 \in \mathcal{M}(X)$ be a probability. A function $k : X \times X \to X$ is called a *Markov–Feller kernel with respect to $\mu_0$* (or, simply, a *Markov–Feller kernel* if there is no danger of confusion) if the following two conditions are satisfied:

(a) $k$ is measurable (that is, $k^{-1}(A)$ belongs to the product $\sigma$-algebra $\mathcal{B}(X) \otimes \mathcal{B}(X)$ whenever $A \in \mathcal{B}(X)$).

(b) For every $f \in C_b(X)$ the function $g_f : X \to \mathbb{R}$ defined by
$g_f(x) = \int f(k(x, y)) \, \mathrm{d}\mu_0(y)$ for every $x \in X$ belongs to $C_b(X)$, as well.

Given a probability $\mu_0 \in \mathcal{M}(X)$ and a Markov–Feller kernel $k$ with respect to $\mu_0$ we define two operators as follows:

$$T : \mathcal{M}(X) \to \mathcal{M}(X) \text{ defined by } T\mu(A) = \int\int 1_A(k(x,y)) \, \mathrm{d}\mu_0(y) \, \mathrm{d}\mu(x)$$

for every $\mu \in \mathcal{M}(X)$ and $A \in \mathcal{B}(X)$

and

$$S : B_b(X) \to B_b(X) \text{ defined by } Sf(x) = \int f(k(x,y)) \, \mathrm{d}\mu_0(y)$$

for every $f \in B_b(X)$ and $x \in X$.

Clearly, $T$ is well-defined (in the sense that $T\mu \in \mathcal{M}(X)$ whenever $\mu \in \mathcal{M}(X)$) and a Markov operator. Using standard facts about functions that are measurable with respect to product $\sigma$-algebras (see, for example, Proposition 5.2.1, p. 159 of Cohn [8]), we obtain that $S$ is well-defined (that is, $Sf \in B_b(X)$ whenever $f \in B_b(X)$).

It is easy to see that

$$\int f(x) \, \mathrm{d}(T\mu)(x) = \int \int f(k(x,y)) \, \mathrm{d}\mu_0(y) \, \mathrm{d}\mu(x). \qquad (2.3.2)$$

(Indeed, the equality is true whenever $f = 1_A$, $A \in \mathcal{B}(X)$, so (2.3.2) is true whenever $f$ is a simple function. If $f \in B_b(X)$, then there exists a sequence $(f_n)_{n \in \mathbb{N}}$ of simple functions such that $(f_n)_{n \in \mathbb{N}}$ converges uniformly to $f$; since $f_n$ satisfies (2.3.2) for every $n \in \mathbb{N}$, it follows that $f$ satisfies (2.3.2), as well.)

Condition (b) in the definition of a Markov–Feller kernel implies that $Sf \in C_b(X)$ whenever $f \in C_b(X)$; thus, the restriction of $S$ to $C_b(X)$ denoted again by $S$) can be thought of as a linear operator from $C_b(X)$ to $C_b(X)$.

Using (2.3.2) we see that $S$ and $T$ satisfy (1.1.1) for every $f \in C_b(X)$ (actually, for every $f \in B_b(X)$) and every $\mu \in \mathcal{M}(X)$; hence, $(S, T)$ is a Markov–Feller pair. We call $(S, T)$ the *Markov–Feller pair induced by $\mu_0$ and $k$*. We say that $(S, T)$ is *induced by a kernel*, if there exist a probability $\mu_0 \in \mathcal{M}(X)$ and a Markov–Feller kernel $k$ with respect to $\mu_0$ such that $(S, T)$ is induced by $\mu_0$ and $k$.

If $(S, T)$ is induced by a kernel, then $T$ is a *Foiaş operator* (for details on Foiaş operators, see Section 12.4 of the book by Lasota and Mackey [36]).

Note that condition (b) in the definition of a Markov–Feller kernel is satisfied whenever the following condition is satisfied:

(b′) For every $x \in X$, every sequence $(x_n)_{n \in \mathbb{N}}$ of elements of $X$ that converges to $x$, every $y \in X$ and every $\varepsilon \in \mathbb{R}$, $\varepsilon > 0$ there exists $n_\varepsilon \in \mathbb{N}$ such that $d(k(x_n, y), k(x, y)) < \varepsilon$ for every $n \geq n_\varepsilon$.

In particular, if $X$ is compact and $k : X \times X \to X$ is continuous (with respect to the product topology on $X \times X$), then $k$ satisfies condition (b′), so $k$ is a Markov–Feller kernel with respect to any probability $\mu_0$, $\mu_0 \in \mathcal{M}(X)$.

If $(S,T)$ is a Markov–Feller pair induced by a continuous function, then $(S,T)$ is also induced by a kernel. Indeed, if $w : X \to X$ is a continuous function such that $(S,T)$ is induced by $w$, and if we define $k : X \times X \to X$ by $k(x,y) = w(x)$ for every $x \in X$ and $y \in X$ (note that $k$ does not depend on $y$), then it is easy to see that $k$ is a Markov–Feller kernel with respect to any probability $\mu_0 \in \mathcal{M}(X)$, and that $(S,T)$ is induced by any probability $\mu_0 \in \mathcal{M}(X)$ and $k$.

We are now in a position to discuss the examples of trivial minimal Markov–Feller pairs that we mentioned earlier.

*Example* 2.3.6. Assume that the Markov–Feller pair $(S,T)$ is induced by a probability $\mu_0$ and by a Markov–Feller kernel $k : X \times X \to X$. For every $x \in X$ let $k_x : X \to X$ be defined by $k_x(y) = k(x,y)$ for every $y \in X$. If supp $\mu_0 = X$ and $k_x$ is a surjective continuous mapping for every $x \in X$, then supp $T\delta_x = X$ for every $x \in X$ (that is, $(S,T)$ is trivially minimal). Indeed, using Fubini's theorem we obtain that

$$T\delta_x(U) = \int\int 1_U(k(z,y))\, d\mu_0(y)\, d\delta_x(z) = \int 1_U(k(x,y))\, d\mu_0(y)$$
$$= \int 1_{k_x^{-1}(U)}(y)\, d\mu_0(y) = \mu_0\left(k_x^{-1}(U)\right) > 0$$

for every $x \in X$ and every open nonempty subset $U$ of $X$.

For example, if $X = \mathbb{R}/\mathbb{Z}$ (the unit circle), if we let $\mu_0$ be the Haar (Lebesgue) measure on $X$, and if $k$ is defined by $k(x,y) = x \oplus y$ for every $x,y \in X$ (as in Example 1.1.11, the sign $\oplus$ stands for the addition modulo 1), then the corresponding Markov–Feller pair $(S,T)$ is trivially minimal. Note that in this case $T\mu = \mu_0$ for every probability $\mu \in \mathcal{M}(X)$, so the range of $T$ is a one-dimensional subspace of $\mathcal{M}(X)$ (the vector subspace spanned by $\mu_0$); accordingly, T is a rank one operator (for details on rank one operators, see [74]).

One can use the unit circle $X = \mathbb{R}/\mathbb{Z}$ to construct other trivially minimal Markov–Feller pairs. In order to describe such an example, we will denote by $\underline{x}$ the first nonnegative number in the equivalence class $x$, $x \in \mathbb{R}/\mathbb{Z}$. Again, let $\mu_0$ be the Haar (Lebesgue) measure on $X$, but let $k$ be defined by $k(x,y) = \sin\left(\frac{\pi}{2}\underline{x}\right) \oplus \sin\left(\frac{\pi}{2}\underline{y}\right)$ for every $x,y \in X$. It is easy to see that the Markov–Feller pair induced by $\mu_0$ and $k$ is trivially minimal. ∎

We have seen in this section that the invariant probabilities of a minimal Markov–Feller pair are supported on the entire space (whenever the Markov–Feller pair has such invariant probabilities, of course); in other words, if a minimal Markov–Feller pair has invariant probabilities, then there exists a unique closed set (the entire space) such that the set is the support of each of the invariant probabilities. The existence of uniquely ergodic Markov–Feller pairs that are not strictly ergodic (like the pair constructed in Example 1.1.14) suggests that some Markov–Feller pairs have the property that they have invariant probabilities, the invariant probabilities of such a pair have the same support, and the common support is different from the entire space.

The above comments suggest two natural questions:

(a) Can we find "formulas" for the support of the (unique) invariant probability of a uniquely ergodic Markov–Feller pair?

(b) If a Markov–Feller pair is not uniquely ergodic, has invariant probabilities, and all these invariant probabilities have the same support, can we find "formulas" for the common support?

The answers to these two questions are among the topics that are discussed in the next chapter. It turns out that the same "formulas" (by a "formula" we mean an equality similar in spirit to the one obtained in Theorem 2.2.2; note that we can already use Theorem 2.2.2 to obtain a "formula" for (a) since the invariant probability of a uniquely ergodic Markov–Feller pair is an ergodic measure; however, in the next chapter we obtain more interesting "formulas") can be used to answer both questions (a) and (b).

# Chapter 3

# Unique Ergodicity of Markov–Feller Operators and Related Topics

In Theorem 2.2.2 we obtained a "formula" for the support of an ergodic measure. Since the (unique) invariant probability of a uniquely ergodic Markov–Feller pair is an ergodic measure, Theorem 2.2.2 can be used to deduce also a "formula" for the support of the invariant probability of a uniquely ergodic Markov–Feller pair. However, the "formula" obtained in this way is generally difficult to apply. That is why, we start this chapter in Section 3.1 by obtaining other formulas that are easier to use. Also in Section 3.1 we show that if a Markov–Feller pair has invariant probabilities, and there exists a closed set $F$ such that $F$ is the support of each invariant probability, then $F$ is given by the same "formulas" that we obtained for the invariant probability of a uniquely ergodic Markov–Feller pair.

In Section 3.2 we define a certain kind of generic points that we call dominant generic points, and we prove that a Markov–Feller pair is uniquely ergodic if and only if there exist dominant generic points defined by the pair (see Theorem 3.2.4).

The structure that the KBBY decomposition generated by a Markov–Feller pair $(S, T)$ defines on the space $(X, d)$ can be compared to that of a fruit. There are several "regions" of $X$ that "protect" the nucleus $\Gamma_1$, the place where the "seeds" (the ergodic measures) are "located." If the fruit is "seedless" ($(S, T)$ does not have invariant measures) there is only one region $\mathcal{D}$ ($= \Omega$ in this case) that protects nothing. Some fruits have only one "seed" (when $(S, T)$ is uniquely ergodic), and some have several "seeds." Many properties of "seeds" of "single-seeded" fruits have corresponding properties for "seeds" of "multi-seeded" fruits. A case in point will be discussed in Section 3.3 where we define the notion of a dominant generic point for a set in order to obtain a necessary and sufficient condition for the ergodicity of an invariant probability, thus obtaining a result for ergodic measures that corresponds to Theorem 3.2.4 (see Theorem 3.3.2).

Also in Section 3.3 we obtain a new proof of Theorem 2.1.3. We believe that our proof is easier to follow, and at the same time offers a characterization of the set $\Gamma_1$ which complements the definition of $\Gamma_1$ as stated by Yosida (see Section 2.1 for Yosida's definition of $\Gamma_1$).

## 3.1  Supports of Invariant Probabilities of Certain Markov–Feller Pairs

Throughout this section we assume given a Markov–Feller pair $(S, T)$ defined on a locally compact separable metric space $(X, d)$.

As pointed out at the end of Section 2.3, one can use Theorem 2.2.2 (which offers a "formula" for the support of an ergodic measure) to find a "formula" for the support of the (unique) invariant probability of $(S, T)$ whenever $(S, T)$ is uniquely ergodic. Indeed, as pointed out in Section 2.1 after Theorem 2.1.3, if $(S, T)$ is uniquely ergodic, then $\Gamma_{cp} = \Gamma_1 = [x]$ for all $x \in \Gamma_1$; if $\mu$ is the invariant probability of $(S, T)$, then Theorem 2.2.2 implies that

$$\operatorname{supp} \mu = \bigcap_{x \in \Gamma_{cp}} \overline{\mathcal{O}(x)}. \tag{3.1.1}$$

The problem with (3.1.1) is that, in general, it is hard to decide if an element $x$ of $X$ belongs to $\Gamma_{cp}$; it is much easier to decide if $x$ belongs to $\Gamma$, to $\Gamma_0$, or even to $\Gamma_c$.

In this section we will show that we can replace $\Gamma_{cp}$ by $\Gamma_c$, $\Gamma_0$, or $\Gamma$ in (3.1.1). We will also prove that if $(S, T)$ is not necessarily uniquely ergodic, but has invariant probabilities, and if there exists a closed subset $F$ of $X$ such that $F$ is the support of any of the invariant probabilities of $(S, T)$, then $F$ is given by the same formulas obtained for the support of the probability of a uniquely ergodic Markov–Feller pair. Finally, we will develop a somewhat "probabilistic" terminology which, we believe, helps in obtaining a better intuitive understanding of these "formulas."

Set $\gamma = \bigcap_{x \in \Gamma} \overline{\mathcal{O}(x)}$, $\gamma_0 = \bigcap_{x \in \Gamma_0} \overline{\mathcal{O}(x)}$, $\gamma_c = \bigcap_{x \in \Gamma_c} \overline{\mathcal{O}(x)}$, and $\gamma_{cp} = \bigcap_{x \in \Gamma_{cp}} \overline{\mathcal{O}(x)}$.

**Theorem 3.1.1.** *Assume that $(S, T)$ is uniquely ergodic. If $\mu$ is the invariant probability of $(S, T)$, then $\operatorname{supp} \mu = \gamma_{cp} = \gamma_c = \gamma_0 = \gamma$.*

*Proof.* Since $\Gamma_{cp} \subseteq \Gamma_c \subseteq \Gamma_0 \subseteq \Gamma$ (the fact that $\Gamma_0 \subseteq \Gamma$ can be deduced using Theorem 1.3.2), it follows that $\gamma_{cp} \supseteq \gamma_c \supseteq \gamma_0 \supseteq \gamma$. Thus, in view of (3.1.1), in order to complete the proof, it is enough to show that

$$\operatorname{supp} \mu \subseteq \gamma. \tag{3.1.2}$$

The proof of (3.1.2) is based upon the same approach as the proof of Theorem 2.2.1-(a) and part (a) of the proof of Theorem 2.2.2.

Clearly, in order to prove (3.1.2), we have to prove that for every $x \in \Gamma$ and every $y \in \operatorname{supp} \mu$ there exists a sequence $(y_n)_{n \in \mathbb{N}}$ of elements of $\mathcal{O}(x)$ such that $(y_n)_{n \in \mathbb{N}}$ converges to $y$.

To this end, let $x \in \Gamma$ and $y \in \operatorname{supp} \mu$.

Since $x \in \Gamma$, there exists a Banach limit $L$ such that $\varepsilon_x^{(L)}$ is an elementary $T$-invariant measure. Since $(S, T)$ is uniquely ergodic, it follows that $\varepsilon_x^{(L)} = a\mu$ for some $a \in \mathbb{R}$, $a > 0$.

Since $X$ is locally compact, there exists $\alpha \in \mathbb{R}$, $\alpha > 0$ such that $\overline{B(y, \alpha)}$ is a compact subset of $X$. Accordingly, for every $n \in \mathbb{N}$ the function $f_n : X \to \mathbb{R}$ defined by $f_n(z) = d\left(z, X \setminus B\left(y, \frac{\alpha}{n}\right)\right)$ for every $z \in X$ is an element of $C_0(X)$; since $f_n(z) > 0$ whenever $z \in B\left(y, \frac{\alpha}{n}\right)$, and since $y \in \operatorname{supp} \mu$, it follows that $0 <$ $\langle f_n, a\mu \rangle = \left\langle f_n, \varepsilon_x^{(L)} \right\rangle = L\left( \left( \langle f_n, T^k \delta_x \rangle \right)_{k \in \mathbb{N} \cup \{0\}} \right)$. Let $k_n$ be the first integer such that $\langle f_n, T^{k_n} \delta_x \rangle > 0$ (obviously, there exists $k_n$ with the stated properties because $\langle f_n, T^k \delta_x \rangle > 0$ for infinitely many $k$'s). Since $z \in B\left(y, \frac{\alpha}{n}\right)$ whenever $f_n(z) > 0$, it follows that $B\left(y, \frac{\alpha}{n}\right) \cap \operatorname{supp} T^{n_k} \delta_x \neq \emptyset$. Let $y_n \in B\left(y, \frac{\alpha}{n}\right) \cap \operatorname{supp} T^{n_k} \delta_x$.

Clearly, $y_n \in \mathcal{O}(x)$ for every $n \in \mathbb{N}$, and the sequence $(y_n)_{n \in \mathbb{N}}$ converges to $y$. $\qquad\square$

**Observation.** Note that in proving the above theorem, we used (3.1.1) in order to conclude that $\gamma \subseteq \operatorname{supp} \mu$; the equality (3.1.1) is proved using Theorem 2.1.3 and Theorem 2.2.2, while the proof of Theorem 2.2.2 uses again Theorem 2.1.3. Thus, the "heavy machinery" of the KBBY decomposition (as discussed in Chapter 13, Section 4 of Yosida [70]) is used twice. Let us discuss briefly another proof of the fact that $\gamma \subseteq \operatorname{supp} \mu$ which does not use Theorem 2.1.3.

Let $f \in C_0(X)$, $f \geq 0$ be such that $\langle f, \mu \rangle > 0$. Using Corollary 1.2.7-(b) we obtain that there exists $x_0 \in \operatorname{supp} \mu$ such that the sequence $\left( \frac{1}{n} \sum_{k=0}^{n-1} S^k f(x_0) \right)_{n \in \mathbb{N}}$ converges, and $\lim_{n \to \infty} \frac{1}{n} \sum_{k=0}^{n-1} S^k f(x_0) > 0$. Accordingly, $x_0 \in \Gamma$, so there exists a Banach limit $L$ such that $\varepsilon_{x_0}^{(L)}$ is an elementary $T$-invariant measure. Since $\mu$ is the unique $T$-invariant probability, it follows that $\varepsilon_{x_0}^{(L)} = a\mu$ for some $a \in \mathbb{R}$, $a > 0$. Using Theorem 2.2.1-(b) we obtain that $\overline{\mathcal{O}(x_0)} = \operatorname{supp} \varepsilon_{x_0}^{(L)} = \operatorname{supp} \mu$. Since $x_0 \in \Gamma$, it follows that $\gamma \subseteq \operatorname{supp} \mu$. $\qquad\blacksquare$

If $(X, d)$ is a compact metric space, and if $(S, T)$ is uniquely ergodic, then the "formulas" of Theorem 3.1.1 can be significantly simplified. More precisely, we have:

**Corollary 3.1.2.** *Assume that $(S, T)$ is a uniquely ergodic Markov–Feller pair, and that $(X, d)$ is a compact metric space. If $\mu$ is the invariant probability of $(S, T)$, then $\operatorname{supp} \mu = \bigcap_{x \in X} \overline{\mathcal{O}(x)}$.*

Note that under the conditions of the corollary, using a well known result (see, for example, Proposition 1.2, p. 178 of Krengel [32]), it follows that $X = \Gamma_{cp}$, so, obviously, Theorem 3.1.1 implies that the assertion of the corollary is true.

Alternatively, we can prove the corollary without using Theorem 3.1.1. Indeed, if $x \in X$, then $\varepsilon_x^{(L)} = \mu$ for every Banach limit $L$, so, supp $\mu \subseteq \overline{\mathcal{O}(x)}$ by Theorem 2.2.1-(a); hence, supp $\mu \subseteq \bigcap_{x \in X} \overline{\mathcal{O}(x)}$; if $x \in$ supp $\mu$ and $L$ is a Banach limit, then, again $\varepsilon_x^{(L)} = \mu$, so Theorem 2.2.1-(b) implies that supp $\mu = \overline{\mathcal{O}(x)}$; therefore, supp $\mu \supseteq \bigcap_{x \in X} \overline{\mathcal{O}(x)}$.

We now turn our discussion to the second topic of this section. The next theorem tells us that if $(S, T)$ has invariant probabilities, and there exists a closed subset $F$ of $X$ such that each invariant probability is supported on $F$, then the same "formulas" that are valid for the support of the invariant probability of a uniquely ergodic Markov–Feller pair, are also valid for $F$.

**Theorem 3.1.3.** *Assume that $(S, T)$ has invariant probabilities, and that there exists a closed subset $F$ of $X$ such that supp $\mu = F$ whenever $\mu$ is a $T$-invariant probability. Then $F = \gamma_{cp} = \gamma_c = \gamma_0 = \gamma$.*

*Proof.* Since $\gamma \subseteq \gamma_0 \subseteq \gamma_c \subseteq \gamma_{cp}$ (as pointed out in the proof of Theorem 3.1.1), it is enough to prove that:

(a) $F \subseteq \gamma$        and

(b) $\gamma_{cp} \subseteq F$.

(a) For every $x \in \Gamma$ there exists a Banach limit $L$ (which, in general, depends on $x$, of course) such that $\varepsilon_x^{(L)}$ is an elementary $T$-invariant measure; by Theorem 2.2.1-(a), $F = $ supp $\varepsilon_x^{(L)} \subseteq \overline{\mathcal{O}(x)}$. Thus, $F \subseteq \gamma$.

(b) Since $(S, T)$ has invariant probabilities, by Theorem 1.2.1, $(S, T)$ has also ergodic measures. By Theorem 2.1.3 there exists an ergodic measure $\varepsilon_{x'}$ for some $x' \in \Gamma_1$ and $\varepsilon_{x'}([x']) = 1$. Since supp $\varepsilon_{x'} = F$, it follows that $F \cap [x'] \neq \emptyset$, so there exists $x \in F \cap [x']$. Since $\varepsilon_x = \varepsilon_{x'}$ is an elementary $T$-invariant measure with respect to $x$ and a suitably chosen Banach limit $L$, Theorem 2.2.1-(b) implies that $F = \overline{\mathcal{O}(x)}$. Since $x \in \Gamma_1 \subseteq \Gamma_{cp}$, we conclude that $\gamma_{cp} \subseteq F$.    □

Following Högnäs and Mukherjea (see p. 180 of [29]) we say that $x$ *leads to* $y$ whenever $y \in \overline{\mathcal{O}(x)}$, $x \in X$, $y \in X$.

Given a nonempty subset $A$ of $X$ and $y \in X$, we say that $y$ is a *universal element with respect to $A$* if $x$ leads to $y$ whenever $x \in A$ (that is, if $y \in \bigcap_{x \in A} \overline{\mathcal{O}(x)}$). We say that $y$ is a *universal element* if $y$ is universal with respect to $X$. The term "universal element" was suggested to us by Furstenberg.

Theorem 3.1.1, Corollary 3.1.2, and Theorem 3.1.3 can be restated in terms of universal elements as follows:

**Theorem 3.1.4.** *If $(S, T)$ is uniquely ergodic and if $\mu$ is the invariant probability of $(S, T)$, then supp $\mu$ is equal to each of the following sets:*

– *the set of all universal elements with respect to $\Gamma_{cp}$;*

    – *the set of all universal elements with respect to* $\Gamma_c$;

    – *the set of all universal elements with respect to* $\Gamma_0$;

    – *the set of all universal elements with respect to* $\Gamma$.

Note that if $(S,T)$ is uniquely ergodic, if $\mu$ is the invariant probability of $(S,T)$, and if there exists a universal element $y$ (with respect to $X$), then $y \in$ supp $\mu$.

**Corollary 3.1.5.** *If $(X,d)$ is a compact metric space, if $(S,T)$ is uniquely ergodic, and if $\mu$ is the invariant probability of $(S,T)$, then supp $\mu$ is the set of all universal elements (generated by $(S,T)$).*

**Theorem 3.1.6.** *If $(S,T)$ has invariant probabilities, and there exists a closed subset $F$ of $X$ such that supp $\mu = F$ whenever $\mu$ is a $T$-invariant probability, then $F$ is equal to each of the following sets:*

    – *the set of all universal elements with respect to* $\Gamma_{cp}$;

    – *the set of all universal elements with respect to* $\Gamma_c$;

    – *the set of all universal elements with respect to* $\Gamma_0$;

    – *the set of all universal elements with respect to* $\Gamma$.

Like in the case of Theorem 3.1.4, if $y$ is a universal element and if $(S,T)$ and $F$ are as in Theorem 3.1.6, then $y \in F$.

## 3.2 Generic Points and Unique Ergodicity

Our goal in this section is to obtain a necessary and sufficient condition for the unique ergodicity of a Markov–Feller pair.

Like in Section 3.1, throughout this section we assume given a Markov–Feller pair $(S,T)$ defined on a locally compact separable metric space $(X,d)$.

We call $x \in X$ a *generic point* (see Furstenberg [20]) if the sequence

$$\left( \frac{1}{n} \sum_{n=1}^{n-1} S^k f(x) \right)_{n \in \mathbb{N}}$$

converges whenever $f \in C_0(X)$. Such points are also called *quasi-regular* (see Oxtoby [53] and Krengel's book [32]).

A generic point $x \in X$ is called *nonsingular* if $\lim_{n \to \infty} \frac{1}{n} \sum_{k=0}^{n-1} S^k f(x) \neq 0$ for some $f \in C_0(X)$ (that is, a generic point is nonsingular if and only if $x \in \Gamma_c$). Naturally, a generic point $x \in X$ is called *singular* if $x$ is not nonsingular, so a generic point is singular if and only if the point belongs to $\mathcal{D}$.

A point $x \in X$ is called a *dominant generic point* if $x$ is a nonsingular generic point and if the following condition (called the *DGP condition*) is satisfied: if

$y \in X$ and $f \in C_0(X)$, $f \geq 0$ are such that the sequence $\left( \frac{1}{n} \sum_{k=0}^{n-1} S^k f(y) \right)_{n \in \mathbb{N}}$ is convergent, then $\lim\limits_{n \to \infty} \frac{1}{n} \sum_{k=0}^{n-1} S^k f(y) \leq \lim\limits_{n \to \infty} \frac{1}{n} \sum_{k=0}^{n-1} S^k f(x)$.

**Proposition 3.2.1.** *If $x \in X$ is a dominant generic point, then $x \in \Gamma_{cp}$.*

*Proof.* Let $\varepsilon_x$ be the measure defined in the subsection *The KBBY Decomposition* of Section 1.2. We have to show that $\|\varepsilon_x\| = 1$.

Since a dominant generic point is nonsingular, it follows that $\varepsilon_x \neq 0$. As pointed out in Section 2.1, $\varepsilon_x$ is an elementary $T$-invariant measure; hence, if we set $\mu = \dfrac{\varepsilon_x}{\|\varepsilon_x\|}$, then $\mu$ is a $T$-invariant probability.

Let $\varepsilon \in \mathbb{R}$, $\varepsilon > 0$. Since $\mu$ is a probability, there exists $f \in C_0(X)$ such that $0 \leq f \leq 1$ and $\langle f, \mu \rangle > 1 - \varepsilon$.

It follows that there exists $y \in X$ such that the sequence $\left( \frac{1}{n} \sum_{k=0}^{n-1} S^k f(y) \right)_{n \in \mathbb{N}}$

converges, and $\lim\limits_{n \to \infty} \frac{1}{n} \sum_{k=0}^{n-1} S^k f(y) > 1 - \varepsilon$ (indeed, assume that there is no such

$y$; by Theorem 1.2.6 the sequence $\left( \frac{1}{n} \sum_{k=0}^{n-1} S^k f \right)_{n \in \mathbb{N}}$ converges $\mu$-a.e. to some

$\mu$-integrable function $g$; since we assume that for every $y \in X$ the sequence $\left( \frac{1}{n} \sum_{k=0}^{n-1} S^k f(y) \right)_{n \in \mathbb{N}}$ either diverges, or else $\lim\limits_{n \to \infty} \frac{1}{n} \sum_{k=0}^{n-1} S^k f(y) \leq 1 - \varepsilon$, we ob-

tain that $\langle f, \mu \rangle = \int g \, d\mu \leq 1 - \varepsilon$; therefore, we obtained a contradiction).

Since $x$ is a dominant generic point, it follows that

$$\lim_{n \to \infty} \frac{1}{n} \sum_{k=0}^{n-1} S^k f(x) \geq \lim_{n \to \infty} \frac{1}{n} \sum_{k=0}^{n-1} S^k f(y) > 1 - \varepsilon;$$

therefore, $\|\varepsilon_x\| > 1 - \varepsilon$.

Since $\|\varepsilon_x\| > 1 - \varepsilon$ for every $\varepsilon \in \mathbb{R}$, $\varepsilon > 0$, and since $0 < \|\varepsilon_x\| \leq 1$, we conclude that $\|\varepsilon_x\| = 1$.                                                                                    $\square$

**Proposition 3.2.2.** *If there exists a dominant generic point defined by $(S, T)$, then $(S, T)$ is uniquely ergodic.*

*Proof.* Let $x$ be a dominant generic point. Then $\varepsilon_x$ is an elementary $T$-invariant measure, and, Proposition 3.2.1 implies that $\varepsilon_x$ is actually a $T$-invariant proba-bility, so $(S, T)$ has invariant probabilities. Thus, in order to complete the proof of the proposition, we have to show that $\varepsilon_x$ is the unique invariant probability of $(S, T)$.

To this end, let $\mu$ be a $T$-invariant probability. We have to prove that $\mu = \varepsilon_x$, but since both $\mu$ and $\varepsilon_x$ are probabilities, it is enough to prove that $\mu \leq \varepsilon_x$; that is, it is enough to prove that $\langle f, \mu \rangle \leq \langle f, \varepsilon_x \rangle$ for every $f \in C_0(X)$, $f \geq 0$.

Thus, let $f \in C_0(X)$, $f \geq 0$. By Theorem 1.2.6 there exists a $\mu$-integrable function $g$, and a subset $A$ of $X$, $A \in \mathcal{B}(X)$ such that $\mu(A) = 1$, the sequence

$$\left( \frac{1}{n} \sum_{k=0}^{n-1} S^k f(y) \right)_{n \in \mathbb{N}}$$ converges to $g(y)$ for every $y \in A$, and such that $\langle f, \mu \rangle = \langle g, \mu \rangle$.

Using the notation $f^*(x) = \lim_{n \to \infty} \frac{1}{n} \sum_{k=0}^{n-1} S^k f(x)$ which was introduced in Section 2.1 before the definition of $\Gamma_1$, and using the fact that $x$ is a dominant generic point, we obtain that $\langle f, \mu \rangle = \langle g, \mu \rangle \leq \int_A f^*(x) \, d\mu(y) = f^*(x) = \langle f, \varepsilon_x \rangle$.  □

Proposition 3.2.2 has a converse which we discuss next.

**Proposition 3.2.3.** *If $(S, T)$ is uniquely ergodic, then there exists a dominant generic point defined by $(S, T)$.*

We will offer two proofs of the proposition: one that uses the KBBY decomposition, and one that does not. The proof that uses the decomposition is shorter, but uses implicitly the entire machinery of the KBBY decomposition (like, for example, the representation of any invariant measure as a convex combination of certain elementary $T$-invariant measures (see formula (14) on p. 395 of Yosida [70])).

*First Proof (Using the KBBY Decomposition).* Since we assume that $(S, T)$ is uniquely ergodic, Theorem 1.2.2 implies that the unique $T$-invariant probability of $(S, T)$ is an ergodic measure. Using Theorem 2.1.3-(a) we obtain that $\Gamma_1 \neq \emptyset$. In order to complete the proof of the proposition we will show that any $x \in \Gamma_1$ is a dominant generic point.

To this end, let $x \in \Gamma_1$. Since $x$ is obviously a nonsingular generic point, in order to prove that $x$ is dominant, it is enough to prove that the DGP condition is satisfied.

Let $y \in X$ and let $f \in C_0(X)$, $f \geq 0$, be such that the sequence

$$\left( \frac{1}{n} \sum_{k=0}^{n-1} S^k f(y) \right)_{n \in \mathbb{N}}$$ converges. Let $L_0$ be a Banach limit, and let $L : l^\infty \to \mathbb{R}$

be defined by $L\left( (a_n)_{n \in \mathbb{N}} \right) = L_0 \left( \left( \frac{1}{n} \sum_{k=1}^{n} a_k \right)_{n \in \mathbb{N}} \right)$ for every $(a_n)_{n \in \mathbb{N}} \in l^\infty$. Then $L$ is a Banach limit (note that this type of Banach limits have been used in Section 2.1 after Theorem 2.1.1). Let $\phi : C_b(X) \to \mathbb{R}$ be defined by $\phi(h) = L\left( (S^n h(y))_{n \in \mathbb{N} \cup \{0\}} \right)$ for every $h \in C_b(X)$. (Note that $\phi$ is well defined since

$(S^n h(y))_{n \in \mathbb{N} \cup \{0\}} \in l^\infty$ for every $h \in C_b(X)$ because $S$ is a positive contraction of $C_b(X)$.)

Clearly, $\phi(1_X) = 1$. Since $\phi(h) = L_0\left(\left(\dfrac{1}{n}\sum_{k=0}^{n-1} S^k h(y)\right)_{n \in \mathbb{N}}\right)$,

$\dfrac{1}{m}\sum_{k=0}^{m-1} S^k h(y) - \dfrac{1}{m}\sum_{k=0}^{m-1} S^{k+1} h(y) = \dfrac{1}{m}\left(h(y) - S^m h(y)\right)$ for every $m \in \mathbb{N}$, and since

$\lim\limits_{n \to \infty} \dfrac{1}{n}\left(h(y) - S^n h(y)\right) = 0$ for every $h \in C_b(X)$, it follows that $\phi(Sh) = \phi(h)$ for every $h \in C_0(X)$ (actually for every $h \in C_b(X)$). Thus, we can apply the Lasota–Yorke lemma (Theorem 1.2.4) in order to obtain that the restriction $\mu_\phi$ of $\phi$ to $C_0(X)$ is a $T$-invariant measure. Since $(S,T)$ is uniquely ergodic and $x \in \Gamma_1$, it follows that $\varepsilon_x$ is the unique $T$-invariant probability, and $\mu_\phi = a\varepsilon_x$ for some $a \in \mathbb{R}$, $0 \le a \le 1$.

We obtain that

$$\lim_{n \to \infty} \dfrac{1}{n}\sum_{k=0}^{n-1} S^k f(y) = L_0\left(\left(\dfrac{1}{n}\sum_{k=0}^{n-1} S^k f(y)\right)_{n \in \mathbb{N}}\right) = L\left((S^n f(y))_{n \in \mathbb{N} \cup \{0\}}\right)$$

$$= \langle f, \mu_\phi \rangle \le \langle f, \varepsilon_x \rangle = \lim_{n \to \infty} \dfrac{1}{n}\sum_{k=0}^{n-1} S^k f(x).$$

Thus, the DGP condition is satisfied, so $x$ is a dominant generic point.  □

*Second Proof (Without Using the KBBY Decomposition).* Assume that $(S,T)$ is uniquely ergodic, and let $\mu$ be the unique $T$-invariant probability.

In order to prove the proposition we will construct a subset $A$ of $X$ such that $A \in \mathcal{B}(X)$, $\mu(A) = 1$, and such that the sequence $\left(\dfrac{1}{n}\sum_{k=0}^{n-1} S^k f(x)\right)_{n \in \mathbb{N}}$ converges for every $x \in A$ and $f \in C_0(X)$; next we conclude the proof of the proposition by showing that each $x \in A$ is a dominant generic point.

Let $f \in C_0(X)$, $f \ge 0$. By Theorem 1.2.6 there exist a subset $A_f^{(1)}$ of $X$ such that $A_f^{(1)} \in \mathcal{B}(X)$, $\mu\left(A_f^{(1)}\right) = 1$, and a $\mu$-integrable function $g_f$ such that the sequence $\left(\dfrac{1}{n}\sum_{k=0}^{n-1} S^k f(x)\right)_{n \in \mathbb{N}}$ converges to $g_f(x)$ for every $x \in A_f^{(1)}$ and such that $\langle f, \mu \rangle = \int g_f \, d\mu$.

We now show that $g_f(x) \le \langle f, \mu \rangle$ for every $x \in A_f^{(1)}$. To this end, let $x \in A_f^{(1)}$. If $g_f(x) < 0$, then, clearly, $g_f(x) \le \langle f, \mu \rangle$ (because we assume that $f \ge 0$, so $\langle f, \mu \rangle \ge 0$); thus, we may and do assume that $g_f(x) \ge 0$. Let $L_0$ be a Banach limit, and let $L : l^\infty \to \mathbb{R}$ be defined by $L\left((a_n)_{n \in \mathbb{N}}\right) = L_0\left(\left(\dfrac{1}{n}\sum_{k=1}^{n} a_k\right)_{n \in \mathbb{N}}\right)$

for every $(a_n)_{n \in \mathbb{N}} \in l^\infty$. Then $L$ is also a Banach limit. Let $\phi : C_b(X) \to \mathbb{R}$ be defined by $\phi(h) = L\left( (S^n h(x))_{n \in \mathbb{N} \cup \{0\}} \right)$ for every $h \in C_b(X)$. Note that we used a similar construction (for a different purpose, however) in the first proof of the proposition. As in the first proof, it follows that $\phi$ is a linear functional, $\phi(1_X) = 1$, and $\phi(Sh) = \phi(h)$ for every $h \in C_0(X)$; hence, we can apply the Lasota–Yorke lemma (Theorem 1.2.4); we obtain that the restriction $\mu_\phi$ of $\phi$ to $C_0(X)$ is a $T$-invariant (positive) measure, and that $\|\mu_\phi\| \le 1$. Since $\mu$ is the only $T$-invariant probability, it follows that $\mu_\phi = a\mu$ for some $a \in \mathbb{R}$, $0 \le a \le 1$. Using the fact that the sequence $\left( \dfrac{1}{n} \displaystyle\sum_{k=0}^{n-1} S^k f(x) \right)_{n \in \mathbb{N}}$ converges to $g_f(x)$, we obtain that

$$g_f(x) = L_0\left( \left( \frac{1}{n} \sum_{k=0}^{n-1} S^k f(x) \right)_{n \in \mathbb{N}} \right) = L\left( (S^n f(x))_{n \in \mathbb{N} \cup \{0\}} \right)$$
$$= \phi(f) = \langle f, \mu_\phi \rangle = \langle f, a\mu \rangle \le \langle f, \mu \rangle.$$

Since $g_f(x) \le \langle f, \mu \rangle$ for every $x \in A_f^{(1)}$, since $\mu\left( A_f^{(1)} \right) = 1$, and since $\int g_f \, d\mu = \langle f, \mu \rangle$, it follows that there exists a subset $A_f$ of $A_f^{(1)}$ such that $A_f \in \mathcal{B}(X)$, $\mu(A_f) = 1$, and $g_f(x) = \langle f, \mu \rangle$ for every $x \in A_f$.

Since each $f \in C_0(X)$ is of the form $f = f^+ - f^-$ where $f^+ = f \vee 0$ and $f^- = (-f) \vee 0$, and since $f^+ \ge 0$, $f^- \ge 0$, $f^+ \in C_0(X)$, $f^- \in C_0(X)$, it follows that for every $f \in C_0(X)$ there exists $A_f \in \mathcal{B}(X)$ such that $\mu(A_f) = 1$, and such that the sequence $\left( \dfrac{1}{n} \displaystyle\sum_{k=0}^{n-1} S^k f(x) \right)_{n \in \mathbb{N}}$ converges to $\langle f, \mu \rangle$ for every $x \in A_f$.

Since $C_0(X)$ is separable (see Theorem 1.3.3) we may and do pick a countable dense subset $\mathcal{H}$ of $C_0(X)$. Let $A = \displaystyle\bigcap_{h \in \mathcal{H}} A_h$. Clearly, $A \in \mathcal{B}(X)$ and $\mu(A) = 1$.

It follows that the sequence $\left( \dfrac{1}{n} \displaystyle\sum_{k=0}^{n-1} S^k f(x) \right)_{n \in \mathbb{N}}$ converges to $\langle f, \mu \rangle$ for every $x \in A$ and $f \in C_0(X)$. Indeed, let $f \in C_0(X)$, let $x \in A$, and let $\varepsilon \in \mathbb{R}$, $\varepsilon > 0$; then, there exists $h \in \mathcal{H}$ such that $\|f - h\| < \dfrac{\varepsilon}{3}$, and there exists $n_\varepsilon \in \mathbb{N}$ such that

$$\left| \frac{1}{n} \sum_{k=0}^{n-1} S^k h(x) - \langle h, \mu \rangle \right| < \frac{\varepsilon}{3} \text{ for every } n \ge n_\varepsilon;$$ since $S$ is a positive contraction of $C_0(X)$, it follows that

$$\left| \frac{1}{n} \sum_{k=0}^{n-1} S^k f(x) - \langle f, \mu \rangle \right| \le \left| \frac{1}{n} \sum_{k=0}^{n-1} S^k (f - h)(x) \right| + \left| \frac{1}{n} \sum_{k=0}^{n-1} S^k h(x) - \langle h, \mu \rangle \right|$$
$$+ |\langle h, \mu \rangle - \langle f, \mu \rangle| < \frac{\varepsilon}{3} + \frac{\varepsilon}{3} + \frac{\varepsilon}{3} = \varepsilon$$

for every $n \in \mathbb{N}$, $n \geq n_\varepsilon$.

It now remains to prove that every $x \in A$ is a dominant generic point, but the proof is based upon the same arguments as the ones that we used to show that the elements of $\Gamma_1$ are dominant generic points in the first proof of the proposition (where, obviously, $A$ plays the role of $\Gamma_1$, and the unique invariant probability is denoted by $\mu$, rather than $\varepsilon_x$). □

Combining Proposition 3.2.2 and Proposition 3.2.3 we obtain:

**Theorem 3.2.4.** *The Markov–Feller pair $(S, T)$ is uniquely ergodic if and only if there exists a dominant generic point defined by $(S, T)$.*

Note that if $(S, T)$ is uniquely ergodic, then the first proof of Proposition 3.2.3 implies that $\Gamma_1$ is the set of all dominant generic points defined by $(S, T)$. If $\mu$ is the unique $T$-invariant probability, then the second proof of Proposition 3.2.3 implies that $\mu(\Gamma_1) = 1$ (the fact that $\mu(\Gamma_1) = 1$ can also be deduced from Chapter 13, Section 4 of Yosida [70]). Using Theorem 3.1.1, we obtain that $\Gamma_1$ is dense in $\gamma$ (in the sense that $\gamma \subseteq \overline{\Gamma_1}$).

**Observation.** Note that Theorem 3.2.4 is relevant only in the noncompact case. If $(X, d)$ is compact, then the theorem is a straightforward consequence of well-known results.

Indeed, if $(X, d)$ is compact, and if $(S, T)$ is uniquely ergodic, then every $x \in X$ is a dominant generic point (see, for example, Proposition 1.2 and Proposition 1.3 on p. 178 of Krengel's book [32]).

Now assume that $(S, T)$ has a dominant generic point, say $x_0 \in X$ (and, of course, assume also that $(X, d)$ is compact). If we assume that $(S, T)$ is not uniquely ergodic, then there exist two distinct $T$-invariant probabilities, say $\mu$ and $\nu$. Since we assume that $\mu \neq \nu$, there exists $f \in C_0(X)$ $(= C_b(X))$, $0 \leq f \leq 1_X$ such that $\langle f, \mu \rangle \neq \langle f, \nu \rangle$. Using Theorem 1.2.6 we obtain that there exist $x_1, x_2 \in X$ such that both sequences $\left( \dfrac{1}{n} \sum_{k=0}^{n-1} S^k f(x_i) \right)_{n \in \mathbb{N}}$, $i = 1, 2$, are convergent, but

$$\lim_{n \to \infty} \frac{1}{n} \sum_{k=0}^{n-1} S^k f(x_1) \neq \lim_{n \to \infty} \frac{1}{n} \sum_{k=0}^{n-1} S^k f(x_2).$$ Since $x_0$ is a dominant generic point,

it follows that $\displaystyle\lim_{n \to \infty} \frac{1}{n} \sum_{k=0}^{n-1} S^k f(x_i) < \lim_{n \to \infty} \frac{1}{n} \sum_{k=0}^{n-1} S^k f(x_0)$ for some $i \in \{1, 2\}$. But

then the sequences $\left( \dfrac{1}{n} \sum_{k=0}^{n-1} S^k (1_X - f)(x_i) \right)_{n \in \mathbb{N}}$ and $\left( \dfrac{1}{n} \sum_{k=0}^{n-1} S^k (1_X - f)(x_0) \right)_{n \in \mathbb{N}}$

converge, and

$$\lim_{n \to \infty} \frac{1}{n} \sum_{k=0}^{n-1} S^k (1_X - f)(x_i) > \lim_{n \to \infty} \frac{1}{n} \sum_{k=0}^{n-1} S^k (1_X - f)(x_0).$$

We have obtained a contradiction since we have assumed that $x_0$ is a dominant generic point. ∎

The Markov–Feller pair $(S, T)$ is called *weak\* mean ergodic* if every $x \in X$ is a generic point. The reason for our terminology lies in the fact that $(S, T)$ is weak\* mean ergodic if and only if the sequence of averages $\left( \frac{1}{n} \sum_{k=0}^{n-1} T^k \mu \right)_{n \in \mathbb{N}}$ converges in the weak\* topology of $\mathcal{M}(X)$ for every $\mu \in \mathcal{M}(X)$. Note that $(S, T)$ is weak\* mean ergodic if and only if $X = \mathcal{D} \cup \Gamma_c$. Although there are Markov–Feller pairs that are not weak\* mean ergodic (see the subsection *The KBBY Decomposition* of Section 1.2), the class of weak\* mean ergodic Markov–Feller pairs is fairly large (see Section 4.2 and Section 4.3). If $(S, T)$ is weak\* mean ergodic, and if $f \in C_0(X)$, then it makes sense to define the function $f^* : X \to \mathbb{R}$,

$$f^*(x) = \lim_{n \to \infty} \frac{1}{n} \sum_{k=0}^{n-1} S^k f(x) \tag{3.2.1}$$

for every $x \in X$.

Given a set $\mathcal{A}$ of real-valued functions defined on $X$, we say that $\mathcal{A}$ has a *common (absolute) maximum* at $x_0 \in X$ if $g(x) \leq g(x_0)$ for every $x \in X$ and $g \in \mathcal{A}$ (that is, $\mathcal{A}$ has a common maximum at $x_0 \in X$ if each $g \in \mathcal{A}$ attains an absolute maximum value at $x_0$).

For weak\* mean ergodic Markov–Feller pairs Theorem 3.2.4 becomes:

**Corollary 3.2.5.** *Assume that $(S, T)$ is a weak\* mean ergodic Markov–Feller pair, and set*

$$\mathcal{A} = \left\{ f^* : X \to \mathbb{R} \ \middle| \ \begin{array}{l} \text{there exists } f \in C_0(X), f \geq 0 \text{ such that } f^* \text{ is} \\ \text{defined by } f \text{ using (3.2.1)} \end{array} \right\}.$$

*Then $(S, T)$ is uniquely ergodic if and only if there exists $x_0 \in X$ such that $\mathcal{A}$ has a common maximum at $x_0$.*

Note that the assertion of Corollary 3.2.5 makes sense since we assume that $(S, T)$ is weak\* mean ergodic, so the set $\mathcal{A}$ as defined in the corollary is nonempty.

If the Markov–Feller pair $(S, T)$ is uniquely ergodic, and $(X, d)$ is a compact metric space, then it is well-known (see, for example, p. 178 of Krengel's book [32]) that the sequence of averages $\left( \frac{1}{n} \sum_{k=0}^{n-1} S^k f \right)_{n \in \mathbb{N}}$ converges not only pointwise, but also uniformly (in the norm topology of $C_0(X) = C_b(X)$) to a constant function whenever $f \in C_0(X)$. By contrast, in the noncompact case, if $(S, T)$ is weak\* mean ergodic and uniquely ergodic, the averages $\left( \frac{1}{n} \sum_{k=0}^{n-1} S^k f \right)_{n \in \mathbb{N}}$, $f \in C_0(X)$ converge pointwise to nonconstant functions, in general. For example, if $(S, T)$ is the Markov–Feller pair of Example 1.1.14, it is easy to see directly that $(S, T)$ is uniquely ergodic and weak* mean ergodic (in Section 4.3 we will prove that if $X$ is discrete (that is, the set $\{x\}$ is open whenever $x \in X$), then any Markov–Feller pair on $(X, d)$ is weak* mean ergodic); let $f = (u_m)_{m \in \mathbb{N}} \in c_0$ be such that $u_1 \neq 0$. Then the sequence $\left( \dfrac{1}{n} \displaystyle\sum_{k=0}^{n-1} S^k f \right)_{n \in \mathbb{N}}$ converges pointwise to $u_1 1_A$ where $A = \{2l - 1 | l \in \mathbb{N}\}$. In general, if $(S, T)$ is a weak* mean ergodic Markov–Feller pair, and if there exists a nonempty Borel subset $A$ of $X$ such that for every $f \in C_0(X)$ there exists $\alpha_f \in \mathbb{R}$ with the property that $f^* = \alpha_f 1_A$ where $f^*$ is given by (3.2.1), and such that the numbers $\alpha_f$, $f \in C_0(X)$ are not all zero, then (by Corollary 3.2.5) $(S, T)$ is uniquely ergodic.

Looking at Example 1.1.14 one may be tempted to believe that if $(S, T)$ is weak* mean ergodic and uniquely ergodic, then the function $f^*$ obtained using (3.2.1) for some $f \in C_0(X)$ is always of the form $a 1_A$ for some $a \in \mathbb{R}$ and $A \in \mathcal{B}(X)$. However, this is not the case. Indeed, let $(S, T)$ be the Markov–Feller pair of Example 1.2.8. Like in Example 1.1.14 it is easy to see directly that $(S, T)$ is uniquely ergodic and weak* mean ergodic. Let $e_1 = (1, 0, 0, \ldots)$. Obviously, $e_1 \in c_0$, and it is not difficult to see that $\displaystyle\lim_{n \to \infty} \frac{1}{n} \sum_{k=0}^{n-1} S^k e_1 = \left(1, \frac{1}{2}, 1, 0, 1, 0, 1, 0, \ldots\right)$.

## 3.3   Generic Points and Ergodic Measures

As pointed out in the outline of topics of the chapter, in this section we offer a necessary and sufficient condition for the ergodicity of an invariant measure (Theorem 3.3.2 below). The condition is similar to the condition for unique ergodicity that was discussed in Theorem 3.2.4. We conclude the section with a new proof of Theorem 2.1.3 which offers a better understanding of the set $\Gamma_1$.

Like in the previous two sections of this chapter, we assume given a Markov–Feller pair $(S, T)$ defined on a locally compact separable metric space $(X, d)$.

We start with a lemma:

**Lemma 3.3.1.** *Let $\mu$ be a $T$-invariant probability measure. Then the following assertions are equivalent:*

(a) *The measure $\mu$ is ergodic.*

(b) *There exists a Borel subset $B$ of $X$ such that $\mu(B) = 1$ and such that the sequence $\left( \dfrac{1}{n} \displaystyle\sum_{k=0}^{n-1} S^k f(x) \right)_{n \in \mathbb{N}}$ converges, and $\displaystyle\lim_{n \to \infty} \frac{1}{n} \sum_{k=0}^{n-1} S^k f(x) = \langle f, \mu \rangle$ for every $x \in B$ and $f \in C_0(X)$.*

*Proof.* (a)$\Rightarrow$(b) By Theorem 1.2.6 and by Lemma 4.2 of Hernández-Lerma and

Lasserre [25], the sequence $\left( \dfrac{1}{n} \displaystyle\sum_{k=0}^{n-1} S^k f \right)_{n \in \mathbb{N}}$ converges $\mu$-a.e. to a function that

is constant $\mu$-a.e. whenever $f \in C_0(X)$.

Since $C_0(X)$ is separable (see Theorem 1.3.3), there exists a countable dense subset $\mathcal{H}$ of $C_0(X)$. In view of our discussion so far, it follows that for every $h \in \mathcal{H}$ there exists a $\mu$-integrable function $g_h \in B_b(X)$ and a subset $B_h$ of $X$ such that

$B_h \in \mathcal{B}(X)$, $\mu(B_h) = 1$, the sequence $\left( \dfrac{1}{n} \displaystyle\sum_{k=0}^{n-1} S^k h(x) \right)_{n \in \mathbb{N}}$ converges to $g_h(x)$ for

every $x \in B_h$, the function $g_h$ is constant on $B_h$, and $g_h(x) = \langle h, \mu \rangle$ for every $x \in B_h$ because $\langle h, \mu \rangle = \langle g_h, \mu \rangle$ by Theorem 1.2.6.

Set $B = \displaystyle\bigcap_{h \in \mathcal{H}} B_h$. Then $B \in \mathcal{B}(X)$ and $\mu(B) = 1$. The arguments used at the

end of the second proof of Proposition 3.2.3 can be applied here, too, in order

to conclude that the sequence $\left( \dfrac{1}{n} \displaystyle\sum_{k=0}^{n-1} S^k f(x) \right)_{n \in \mathbb{N}}$ converges to $\langle f, \mu \rangle$ for every

$f \in C_0(X)$ and $x \in B$.

(b)$\Rightarrow$(a) Assume that $\mu$ is not ergodic, and let $P$ be the transition probability that generates the Markov–Feller pair $(S, T)$. Then there exists a $P$-invariant set $\Theta \in \mathcal{B}(X)$ such that $0 < \mu(\Theta) < 1$.

Let $\nu_i : \mathcal{B}(X) \to \mathbb{R}$, $i = 1, 2$ be two maps defined as follows: $\nu_1(A) = \mu(A \cap \Theta)$ and $\nu_2(A) = \mu(A) - \nu_1(A)$ for every $A \in \mathcal{B}(X)$. Clearly, $\nu_i \in \mathcal{M}(X)$ for every $i = 1, 2$. Since $0 \leq \nu_1 \leq \mu$, it follows that $\nu_2 \geq 0$. Obviously, $\nu_i \neq 0$, $i = 1, 2$ because $\nu_1(\Theta) = \mu(\Theta) > 0$ and $\nu_2(X \setminus \Theta) = \mu(X \setminus \Theta) > 0$.

We now note that $\nu_1$ is $T$-invariant. Indeed, it is easy to see that $\nu_1$ is absolutely continuous with respect to $\mu$, and that $\nu_1 = 1_{\Theta}\mu$ since $\nu_1(A) = \mu(A \cap \Theta) = \int 1_A \cdot 1_{\Theta} \, d\mu$ for every $A \in \mathcal{B}(X)$; taking into consideration that $P(x, \cdot)$ is a probability, and that $P(x, \Theta) = 1$ for every $x \in \Theta$, we obtain that $P(x, A \cap (X \setminus \Theta)) = 0$ whenever $x \in \Theta$, and, consequently,

$$P(x, A \cap (X \setminus \Theta)) \cdot 1_{\Theta}(x) = 0 \tag{3.3.1}$$

for every $x \in X$ and $A \in \mathcal{B}(X)$; using (3.3.1) we obtain that

$$
\begin{aligned}
T\nu_1(A) &= \int P(x, A) \, d\nu_1(x) = \int P(x, A) 1_{\Theta} \, d\mu(x) \\
&= \int (P(x, A \cap \Theta) + P(x, A \cap (X \setminus \Theta))) 1_{\Theta} \, d\mu(x) \\
&= \int P(x, A \cap \Theta) 1_{\Theta}(x) \, d\mu(x) \leq \int P(x, A \cap \Theta) \, d\mu(x) = \mu(A \cap \Theta) \\
&= \nu_1(A)
\end{aligned}
$$

for every $A \in \mathcal{B}(X)$; that is, $T\nu_1 \leq \nu_1$; since $T$ is a Markov operator, it follows that $T\nu_1 = \nu_1$.

Clearly, $\nu_2$ is also $T$-invariant since $\nu_2 = \mu - \nu_1$, and both $\mu$ and $\nu_1$ are $T$-invariant.

Let $\mu_1 = \dfrac{\nu_1}{\|\nu_1\|}$ and $\mu_2 = \dfrac{\nu_2}{\|\nu_2\|}$. Then both $\mu_1$ and $\mu_2$ are $T$-invariant probabilities. Also, $\mu_1 \neq \mu_2$ because $\nu_1(\Theta) \neq 0$ while $\nu_2(\Theta) = 0$. Accordingly, there exists $f \in C_0(X)$, $f \geq 0$ such that $\langle f, \mu_1 \rangle \neq \langle f, \mu_2 \rangle$.

Let $B$ be the set whose existence is assured by our assumption that (b) holds true. Thus, $\mu(B) = 1$, and the sequence $\left( \dfrac{1}{n} \sum_{k=0}^{n-1} S^k f(x) \right)_{n \in \mathbb{N}}$ converges to $\langle f, \mu \rangle$ for every $x \in B$.

Since $\nu_1(\Theta \cap B) = \mu(\Theta \cap B) = \nu_1(\Theta)$, it follows that $\mu_1(\Theta \cap B) = 1$, and since the sequence $\left( \dfrac{1}{n} \sum_{k=0}^{n-1} S^k f(x) \right)_{n \in \mathbb{N}}$ converges to $\langle f, \mu \rangle$ for every $x \in \Theta \cap B$, by Theorem 1.2.6 we obtain that $\langle f, \mu_1 \rangle = \langle f, \mu \rangle$.

Similarly, $\nu_2(B \backslash \Theta) = \mu(B \backslash \Theta) - \nu_1(B \backslash \Theta) = \mu(X \backslash \Theta) - \nu_1(B \backslash \Theta) = \nu_2(X \backslash \Theta)$ implies that $\nu_2(B \backslash \Theta) = 1$, so, since the sequence $\left( \dfrac{1}{n} \sum_{k=0}^{n-1} S^k f(x) \right)_{n \in \mathbb{N}}$ converges to $\langle f, \mu \rangle$ for every $x \in B \backslash \Theta$, using Theorem 1.2.6 we obtain that $\langle f, \mu_2 \rangle = \langle f, \mu \rangle$.

Accordingly, $\langle f, \mu_1 \rangle = \langle f, \mu_2 \rangle = \langle f, \mu \rangle$; that is, we have obtained a contradiction which stems from our assumption that $\mu$ is not ergodic. $\qquad \square$

Let $A \in \mathcal{B}(X)$, $A \neq \emptyset$. As in the case of dominant generic points, we say that $x_0 \in A$ is a *dominant generic point for $A$* if $x_0$ is a nonsingular generic point, and if the following condition (called the *DGP-A condition*) is satisfied: if $x \in A$ and $f \in C_0(X)$, $f \geq 0$ are such that the sequence $\left( \dfrac{1}{n} \sum_{k=0}^{n-1} S^k f(x) \right)_{n \in \mathbb{N}}$ converges, then

$$\lim_{n \to \infty} \frac{1}{n} \sum_{k=0}^{n-1} S^k f(x) \leq \lim_{n \to \infty} \frac{1}{n} \sum_{k=0}^{n-1} S^k f(x_0).$$

We are now in a position to discuss the condition for the ergodicity of an invariant measure that we mentioned earlier. Note the similarity between the next theorem and Theorem 3.2.4.

**Theorem 3.3.2.** *Let $\mu$ be a $T$-invariant probability. Then the following assertions are equivalent:*

(a) *The measure $\mu$ is ergodic.*

(b) *There exists a Borel subset $A$ of $X$ and $x_0 \in A$ such that $\mu(A) = 1$ and $x_0$ is a dominant generic point for $A$.*

*Proof.* (a) $\Rightarrow$ (b) is a straightforward consequence of Lemma 3.3.1 because the lemma implies that if $\mu$ is ergodic, then there exists a Borel subset $A$ of $X$ such that $\mu(A) = 1$, and every $x_0 \in A$ is a dominant generic point for $A$.

(b) $\Rightarrow$ (a) Let $A$ be a Borel subset of $X$ such that $\mu(A) = 1$ and such that there exists a dominant generic point $x_0 \in A$ for $A$ (the existence of $A$ and $x_0$ is assured by (b)).

For every $f \in C_0(X)$ set $B_f = \left\{ x \in X \mid \lim_{n\to\infty} \frac{1}{n} \sum_{k=0}^{n-1} S^k f(x) \text{ exists} \right\}$, $A_f = A \cap B_f$, and let $f^* : X \to \mathbb{R}$ be defined by

$$f^*(x) = \begin{cases} \lim_{n\to\infty} \dfrac{1}{n} \sum_{k=0}^{n-1} S^k f(x) & \text{if } x \in A_f \\ 0 & \text{if } x \notin A_f \end{cases}.$$

Using Theorem 1.2.6 we obtain that $\mu(B_f) = 1$; therefore, $\mu(A_f) = 1$. Thus, $f^*$ is a $\mu$-a.e. limit of the sequence $\left( \dfrac{1}{n} \sum_{k=0}^{n-1} S^k f \right)_{n\in\mathbb{N}}$. Using again Theorem 1.2.6 and standard facts of measure theory, we obtain that $f^*$ is $\mu$-integrable, and that $\langle f, \mu \rangle = \int f^* \, d\mu$.

We now prove that

$$\lim_{n\to\infty} \frac{1}{n} \sum_{k=0}^{n-1} S^k f(x_0) = \langle f, \mu \rangle \tag{3.3.2}$$

for every $f \in C_0(X)$.

To this end, note that the map $\varepsilon_{x_0} : C_0(X) \to \mathbb{R}$ defined by $\varepsilon_{x_0}(f) = \lim_{n\to\infty} \frac{1}{n} \sum_{k=0}^{n-1} S^k f(x_0)$ for every $f \in C_0(X)$ is a positive linear functional, so we may and do think of $\varepsilon_{x_0}$ as an element of $\mathcal{M}(X)$; moreover, the discussion preceding Theorem 2.1.2 in Section 2.1 shows that $\varepsilon_{x_0}$ is $T$-invariant.

Since $x_0$ is a dominant generic point for $A$, we obtain that

$$\langle f, \mu \rangle = \int f^* \, d\mu = \int_{A_f} f^* \, d\mu \le f^*(x_0) \mu(A_f) = \langle f, \varepsilon_{x_0} \rangle$$

for every $f \in C_0(X)$, $f \ge 0$; thus, $\mu \le \varepsilon_{x_0}$.

Since $\mu$ is a probability measure, and $0 \le \|\varepsilon_{x_0}\| \le 1$, it follows that $\mu = \varepsilon_{x_0}$. Thus, (3.3.2) holds true for every $f \in C_0(X)$.

We now use (3.3.2) in order to prove that $\mu$ is an ergodic measure. Clearly, we have to prove that if $B$ is a $T$-invariant Borel subset of $X$ such that $\mu(B) > 0$, then $\mu(B) = 1$.

Thus, let $B \in \mathcal{B}(X)$, $\mu(B) > 0$, and assume that $B$ is $T$-invariant.

Let $\nu \in \mathcal{M}(X)$ be defined by $\nu(E) = \dfrac{\mu(E \cap B)}{\mu(B)}$ for every $E \in \mathcal{B}(X)$. It is easy to see that $\nu$ is absolutely continuous with respect to $\mu$, and that $\nu = \dfrac{1_B}{\mu(B)} \mu$.

In a similar way as in the proof that $\nu_1$ is $T$-invariant in the implication (b)$\Rightarrow$(a)
of Lemma 3.3.1 one can show that $\nu$ is $T$-invariant, too. (Note that here we use
the fact that $B$ is $T$-invariant.)

Since $\nu(B) = 1$, in order to prove that $\mu(B) = 1$, it is enough to prove
that $\mu = \nu$. To this end, let $f \in C_0(X)$, $f \geq 0$. Since $\nu$ is a $T$-invariant element
of $\mathcal{M}(X)$, and a probability, Theorem 1.2.6 implies that $\nu(B_f) = 1$, that the

sequence $\left(\dfrac{1}{n}\sum\limits_{k=0}^{n-1} S^k f\right)_{n\in\mathbb{N}}$ converges $\nu$-a.e. to $f^*$, that $f^*$ is $\nu$-integrable, and

that $\langle f, \nu\rangle = \int f^* \, d\nu$.

Since $\mu$ is a probability, and $\mu(A) = 1$, it follows that

$$\nu(A) = \frac{\mu(A\cap B)}{\mu(B)} = \frac{\mu(X\cap B)}{\mu(B)} - \frac{\mu((X\setminus A)\cap B)}{\mu(B)} = 1.$$

Thus, $\nu(A_f) = \nu(A\cap B_f) = 1$.

Using the fact that $x_0$ is a dominant generic point for $A$, and the equality
(3.3.2), we obtain that

$$\langle f, \nu\rangle = \int f^* \, d\nu = \int_{A_f} f^* \, d\nu \leq \nu(A_f)\lim_{n\to\infty}\frac{1}{n}\sum_{k=0}^{n-1} S^k f(x_0) = \langle f, \mu\rangle. \qquad (3.3.3)$$

Since (3.3.3) holds true for every $f \in C_0(X)$, $f \geq 0$, it follows that $\nu \leq \mu$.
Since both $\mu$ and $\nu$ are probabilities, we conclude that $\mu = \nu$; hence, $\mu(B) = 1$.
Consequently, $\mu$ is an ergodic measure.                                                      $\square$

Theorem 3.3.2 can be used to obtain a new proof of half of Theorem 1.2.2.
More precisely, we have the following corollary:

**Corollary 3.3.3.** *If $(S,T)$ is uniquely ergodic, then the (unique) invariant probability of $(S,T)$ is an ergodic measure.*

*Proof.* If $(S,T)$ is uniquely ergodic, then Theorem 3.2.4 implies that there exists
a dominant generic point, say $x_0 \in X$, defined by $(S,T)$. Thus, (b) of Theorem
3.3.2 is satisfied for $A = X$.                                                                  $\square$

Note that Corollary 3.3.3 can also be proved using Theorem 2.1.3 (in this
case we have to use certain arguments of Section 4 of Chapter 13 of Yosida [70]).

Our goal now is to offer a new proof of Theorem 2.1.3.

To this end, let $\Gamma_c$ be the subset of $X$ defined in the subsection *The KBBY
Decomposition* of Section 1.2, and note that the equivalence relation $\sim$ defined
on $\Gamma_1$ in Section 2.1 can be extended to $\Gamma_c$ in an obvious way: if $x, y \in \Gamma_c$, then

$$x \sim y \text{ if and only if (by definition) } \lim_{n\to\infty}\frac{1}{n}\sum_{k=0}^{n-1} S^k f(x) = \lim_{n\to\infty}\frac{1}{n}\sum_{k=0}^{n-1} S^k f(y) \text{ for}$$

every $f \in C_0(X)$. As in Section 2.1 we denote by $[x]$ the equivalence class of $x$
with respect to $\sim$ whenever $x \in \Gamma_c$.

**Lemma 3.3.4.** *If $x \in \Gamma_c$, then the equivalence class $[x]$ as a subset of $X$ belongs to* $\mathcal{B}(X)$.

*Proof.* Since $C_0(X)$ is a separable Banach space (see Theorem 1.3.3), using arguments similar to the ones used at the end of the second proof of Proposition 3.2.3, we obtain that $\mathcal{D} \in \mathcal{B}(X)$ and $\mathcal{D} \cup \Gamma_c \in \mathcal{B}(X)$; hence, $\Gamma_c \in \mathcal{B}(X)$. Also, the separability of $C_0(X)$ implies that there exists a countable dense subset $\mathcal{H}$ of $C_0(X)$.

Now let $x \in \Gamma_c$, and set

$$A_x = \left\{ y \in \Gamma_c \,\middle|\, \lim_{n \to \infty} \frac{1}{n} \sum_{k=0}^{n-1} S^k h(x) = \lim_{n \to \infty} \frac{1}{n} \sum_{k=0}^{n-1} S^k h(y) \text{ for every } h \in \mathcal{H} \right\}.$$

Since $\left\{ y \in \Gamma_c \,\middle|\, \lim_{n \to \infty} \frac{1}{n} \sum_{k=0}^{n-1} S^k h(x) = \lim_{n \to \infty} \frac{1}{n} \sum_{k=0}^{n-1} S^k h(y) \right\} \in \mathcal{B}(X)$ whenever $h \in C_0(X)$, since $\mathcal{H}$ is a countable subset of $C_0(X)$, and since

$$A_x = \bigcap_{h \in \mathcal{H}} \left\{ y \in \Gamma_c \,\middle|\, \lim_{n \to \infty} \frac{1}{n} \sum_{k=0}^{n-1} S^k h(x) = \lim_{n \to \infty} \frac{1}{n} \sum_{k=0}^{n-1} S^k h(y) \right\},$$

it follows that $A_x \in \mathcal{B}(X)$.

It is easy to see that $A_x \supseteq [x]$. Using arguments similar to the ones used at the end of the second proof of Proposition 3.2.3, we obtain that $A_x \subseteq [x]$. Thus, $[x] = A_x$; so, $[x] \in \mathcal{B}(X)$. $\qquad\square$

Note that if $x \in \Gamma_{cp}$, then $[x] \subseteq \Gamma_{cp}$.

**Theorem 3.3.5.** *Let $\mu$ be a $T$-invariant element of $\mathcal{M}(X)$. Then the following assertions are equivalent:*

(a) *$\mu$ is an ergodic measure.*

(b) *There exists $x \in \Gamma_c$ such that $\mu = \varepsilon_x$ and $\varepsilon_x([x]) = 1$.*

(c) *There exists $x \in \Gamma_{cp}$ such that $\mu = \varepsilon_x$ and $\varepsilon_x([x]) = 1$.*

*Proof.* (a) $\Rightarrow$ (b). Since we assume that $\mu$ is an ergodic measure, Lemma 3.3.1 implies that there exists a Borel subset $B$ of $X$ such that $\mu(B) = 1$, $B \subseteq \Gamma_{cp} \subseteq \Gamma_c$, and such that $\mu = \varepsilon_x$ for every $x \in B$ (note that since $\mu(B) = 1$, it follows that $B$ is nonempty; thus, there exists $x \in \Gamma_c$ such that $\mu = \varepsilon_x$). Given $x \in B$, it follows that $x \sim y$ whenever $y \in B$; therefore, $B \subseteq [x]$; since $\mu(B) = 1$, and since $[x] \in \mathcal{B}(X)$, it follows that $\varepsilon_x([x]) = 1$.

(b) $\Rightarrow$ (c) If $\mu = \varepsilon_x$ where $x \in \Gamma_c$ and $\varepsilon_x([x]) = 1$, then $\varepsilon_x$ is a probability measure, so $x \in \Gamma_{cp}$.

(c) $\Rightarrow$ (b) is obvious since $\Gamma_{cp} \subseteq \Gamma_c$.

(b) $\Rightarrow$ (a) If $\mu = \varepsilon_x$ for some $x \in \Gamma_c$, and $\varepsilon_x([x]) = 1$, then $\mu$ is a probability measure. If we set $B = [x]$, then $\mu$ and $B$ satisfy assertion (b) of Lemma 3.3.1; accordingly, the measure $\mu$ is ergodic.                                                      □

Theorem 3.3.5 allows us to obtain the new proof of Theorem 2.1.3 that we promised earlier.

Let $\mathcal{E} = \{x \in \Gamma_{cp} | \varepsilon_x$ is an ergodic measure$\}$. In view of Theorem 3.3.5, in order to prove Theorem 2.1.3, it is enough to prove that $\Gamma_1 = \mathcal{E}$; that is, it is enough to prove that

(a) $\Gamma_1 \subseteq \mathcal{E}$      and that

(b) $\Gamma_1 \supseteq \mathcal{E}$.

(a) Let $x \in \Gamma_1$. Since $C_0(X)$ is separable, we may and do pick a countable dense subset $\mathcal{H}$ of $C_0(X)$.

Since $x \in \Gamma_1$, it follows that $\varepsilon_x(A_x^{(h)}) = 1$ where

$$A_x^{(h)} = \left\{ y \in X \,\middle|\, \text{the sequence } \left( \frac{1}{n} \sum_{k=0}^{n-1} S^k h(y) \right)_{n \in \mathbb{N}} \right.$$

$$\left. \text{converges to } \lim_{n \to \infty} \frac{1}{n} \sum_{k=0}^{n-1} S^k h(x) \right\}$$

for every $h \in \mathcal{H}$. Thus, $\varepsilon_x(A_x) = 1$ where $A_x = \bigcap_{h \in \mathcal{H}} A_x^{(h)}$. As in the proof of Lemma 3.3.4, it follows that $A_x = [x]$, so by Theorem 3.3.5 we obtain that $x \in \mathcal{E}$.

(b) Let $x \in \mathcal{E}$. Since $\varepsilon_x$ is an ergodic measure, Theorem 3.3.5 implies that $\varepsilon_x([x]) = 1$. Taking into consideration the way in which the set $\Gamma_1$ has been defined (in Section 2.1), we obviously obtain that $x \in \Gamma_1$.

Note that our proof of Theorem 2.1.3 given here shows that the equivalence relation $\sim$ extended to $\Gamma_c$ has the property that if $x \in \Gamma_c$, then $[x] \cap \Gamma_1 = \emptyset$, or else, $[x] \subseteq \Gamma_1$; thus, we obtain the following characterization of $\Gamma_1$:

$$\Gamma_1 = \bigcup_{x \in \mathcal{E}} [x].$$

# Chapter 4

# Equicontinuity

Theorem 3.1.1 implies that if a Markov–Feller pair $(S, T)$ defined on $(X, d)$ is uniquely ergodic, then the sets $\gamma_{cp}$, $\gamma_c$, $\gamma_0$, and $\gamma$ (defined by $(S, T)$, of course) are nonempty (and each of them is equal to the support of the unique $T$-invariant probability). So, a natural question is whether or not the nonemptyness of any of the sets $\gamma_{cp}$, $\gamma_c$, $\gamma_0$, or $\gamma$ implies the unique ergodicity of $(S, T)$. Since the results of Keynes and Newton [31] and of Keane [30] can be used to show that even if the sets $\gamma_{cp}$, $\gamma_c$, $\gamma_0$, and $\gamma$ are nonempty, and each of them is equal to $X$, the Markov–Feller pair may well have several distinct invariant probabilities (see Section 2.3), the next natural question is: is it possible to exhibit a large enough class of Markov–Feller pairs which has the property that if $(S, T)$ is in the class, and at least one of the sets $\gamma_{cp}$, $\gamma_c$, $\gamma_0$, or $\gamma$ is nonempty, then $(S, T)$ is uniquely ergodic? In this chapter we introduce (in Section 4.1) such a class, the $C_0(X)$-equicontinuous Markov–Feller pairs, and we prove (also in Section 4.1 (see Theorem 4.1.8)) that if $(S, T)$ is $C_0(X)$-equicontinuous and any of the four sets $\gamma_{cp}$, $\gamma_c$, $\gamma_0$, or $\gamma$ is nonempty, then $(S, T)$ is uniquely ergodic. Apart for being of intrinsic interest, we believe that the result (Theorem 4.1.8) can be used in applications involving Markov–Feller pairs defined by iterated function systems or by convolutions.

In Section 4.2 we obtain several preliminary results needed to prove (in Section 4.3) a weak* mean ergodic theorem (Theorem 4.3.1) and a pointwise mean ergodic theorem (Corollary 4.3.2) for $C_0(X)$-equicontinuous Markov–Feller pairs.

We conclude Section 4.3 by discussing several examples involving the mean ergodic theorems, a result (Proposition 4.3.6) that complements the mean ergodic theorems, and a discussion of weak* uniquely mean ergodic Markov–Feller pairs (these pairs are related to the mean ergodic theorems and their properties have been the starting point for the results obtained in this work).

## 4.1   Unique Ergodicity and Equicontinuity

As stated in the outline of topics of this chapter, we start the section by defining (and studying the basic properties of) what we call $C_0(X)$-equicontinuous Markov–Feller pairs. Even though the $C_0(X)$-equicontinuity is the most general straightforward extension of M. Rosenblatt's notion of equicontinuous Markov operators (see [59]), it turns out that the $C_0(X)$-equicontinuous Markov–Feller pairs have all the properties that are of interest to us in this chapter. The main result of this section complements Theorem 3.1.1; that is, we prove that if $(S, T)$ is a $C_0(X)$-equicontinuous Markov–Feller pair which has invariant probabilities, and if at least one of the sets $\gamma_{cp}$, $\gamma_c$, $\gamma_0$, or $\gamma$ is nonempty, then $(S, T)$ is uniquely ergodic.

Throughout this section we assume given a Markov–Feller pair $(S, T)$ defined on a locally compact separable metric space $(X, d)$.

We say that $(S, T)$ (or $S$) is $C_0(X)$-*equicontinuous* if the sequence $(S^n f)_{n \in \mathbb{N}}$ is equicontinuous whenever $f \in C_0(X)$. (For the definition of the equicontinuity of a sequence of functions, see the subsection *Equicontinuity* of Section 1.3.)

The notion of equicontinuity of Markov–Feller operators defined on compact spaces in the spirit of this book was introduced by M. Rosenblatt in his pioneering work [59]. Note that the definition of an equicontinuous family (or sequence) of functions stated in the subsection *Equicontinuity* of Section 1.3 is the same as the one stated in Chapter 4, Section 6 of Dunford and Schwartz [15] (in the case of locally compact separable metric spaces, of course); thus, the $C_0(X)$-equicontinuity agrees with (is the same as) Rosenblatt's equicontinuity in the case of a compact metric space.

In 1988 Barnsley, Demko, Elton, and Geronimo (see [4]) introduced another notion of equicontinuity which we call uniform $C_0(X)$-equicontinuity (the Markov–Feller pair $(S, T)$ (or $S$) is *uniform $C_0(X)$-equicontinuous* if the sequence $(S^n f)_{n \in \mathbb{N} \cup \{0\}}$ is uniformly equicontinuous (on $X$) for every $f \in C_0(X)$). In view of Lemma 1.3.6 it is easy to see that (like the $C_0(X)$-equicontinuity), the uniform $C_0(X)$-equicontinuity agrees with Rosenblatt's equicontinuity when dealing with a compact metric space.

In our paper [73] we defined a notion of equicontinuity which might be called uniform $C_0(X)$-equicontinuity on compact subsets; more precisely, we say that $(S, T)$ (or $S$) is *uniformly $C_0(X)$-equicontinuous on the compact subsets of $X$* if the sequence $(S^n f)_{n \in \mathbb{N} \cup \{0\}}$ is uniformly equicontinuous on the compact subsets of $X$ whenever $f \in C_0(X)$. Using Proposition 1.3.7 it turns out that the definitions of $C_0(X)$-equicontinuity and of uniform $C_0(X)$-equicontinuity on compact subsets are equivalent.

One can replace $C_0(X)$ by $C_b(X)$ in the definition of $C_0(X)$-equicontinuity. Thus, we say that $(S, T)$ (or $S$) is *$C_b(X)$-equicontinuous* if the sequence $(S^n f)_{n \in \mathbb{N} \cup \{0\}}$ is equicontinuous whenever $f \in C_b(X)$. Like the $C_0(X)$-equicontinuity and the uniform $C_0(X)$-equicontinuity, the $C_b(X)$-equicontinuity agrees with Rosenblatt's equicontinuity whenever the space under consideration is compact.

Of course, we can also define the uniform $C_b(X)$-equicontinuity on compact subsets (we say that $(S, T)$ (or $S$) is *uniformly $C_b(X)$-equicontinuous on the compact subsets of $X$* if $(S^n f)_{n \in \mathbb{N} \cup \{0\}}$ is uniformly equicontinuous on the compact subsets of $X$ for every $f \in C_b(X)$); however, as in the case of $C_0(X)$-equicontinuity, using Proposition 1.3.7 we obtain that the definitions of $C_b(X)$-equicontinuity and of uniform $C_b(X)$-equicontinuity on compact subsets are equivalent.

If a sequence $(f_n)_{n \in \mathbb{N} \cup \{0\}}$ of real-valued functions on $X$ is uniformly equicontinuous (on $X$), then, for every $n \in \mathbb{N}$, the function $f_n$ is uniformly continuous on $X$; on the other hand, in most noncompact cases $C_b(X)$ contains also functions that are not uniformly continuous; thus, for such a noncompact (locally compact separable metric) space $(X, d)$, if we define the uniform $C_b(X)$-equicontinuity in the obvious way, it turns out that there is no uniformly $C_b(X)$-equicontinuous Markov–Feller pair defined on $(X, d)$; hence, the notion of uniform $C_b(X)$-equicontinuity is not of much interest.

For our discussion here, and for future use we need the following simple lemma.

**Lemma 4.1.1.** *Let $\mathcal{H}$ be a dense subset of $C_0(X)$, and assume that $(S^n g)_{n \in \mathbb{N} \cup \{0\}}$ is equicontinuous for every $g \in \mathcal{H}$. Then $S$ is $C_0(X)$-equicontinuous.*

*Proof.* We have to prove that $(S^n f)_{n \in \mathbb{N} \cup \{0\}}$ is equicontinuous for every $f \in C_0(X)$. Thus, let $f \in C_0(X)$, let $\varepsilon \in \mathbb{R}$, $\varepsilon > 0$, let $(x_k)_{k \in \mathbb{N}}$ be a convergent sequence of elements of $X$, and set $x = \lim_{k \to \infty} x_k$.

Since $\mathcal{H}$ is dense in $C_0(X)$, there exists $g \in \mathcal{H}$ such that $\|f - g\| < \dfrac{\varepsilon}{3}$. Since $g \in \mathcal{H}$, it follows that the sequence $(S^n g)_{n \in \mathbb{N} \cup \{0\}}$ is equicontinuous; thus, there exists $k_\varepsilon \in \mathbb{N}$ such that $|S^n g(x_k) - S^n g(x)| < \dfrac{\varepsilon}{3}$ for every $k \geq k_\varepsilon$ and $n \in \mathbb{N} \cup \{0\}$.

Since $S$ is a positive contraction of $C_0(X)$, it follows that

$$|S^n f(y) - S^n g(y)| = |S^n(f - g)|(y) \leq S^n(|f - g|)(y) \leq \|f - g\| < \frac{\varepsilon}{3}$$

for every $n \in \mathbb{N} \cup \{0\}$ and $y \in X$.

We obtain that

$$|S^n f(x_k) - S^n f(x)| \leq |S^n f(x_k) - S^n g(x_k)| + |S^n g(x_k) - S^n g(x)|$$

$$+ |S^n g(x) - S^n f(x)| < \frac{\varepsilon}{3} + \frac{\varepsilon}{3} + \frac{\varepsilon}{3} = \varepsilon$$

for every $k \geq k_\varepsilon$ and $n \in \mathbb{N} \cup \{0\}$.

Thus, $(S^n f)_{n \in \mathbb{N} \cup \{0\}}$ is equicontinuous. $\square$

Clearly, if $(S, T)$ is uniformly $C_0(X)$-equicontinuous or $C_b(X)$-equicontinuous, then $(S, T)$ is $C_0(X)$-equicontinuous. The next examples show that, in general, a $C_0(X)$-equicontinuous Markov–Feller pair does not have to be uniformly $C_0(X)$-equicontinuous or $C_b(X)$-equicontinuous.

*Example* 4.1.2. Let $X$ be the interval $[0,\infty)$ endowed with the usual distance defined by the absolute value, let $\psi : [0,\infty) \to \mathbb{R}$ be defined by $\psi(x) = \dfrac{1 + \sin(x^2)}{2}$ for every $x \in [0,\infty)$, and consider the map $P : X \times \mathcal{B}(X) \to \mathbb{R}$ defined by

$$
P(x, A) = \begin{cases}
0 & \text{if } \left\{\sqrt{\frac{\pi}{2}}, \sqrt{\frac{3\pi}{2}}\right\} \cap A = \emptyset \\
\psi(x) & \text{if } \sqrt{\frac{\pi}{2}} \in A \text{ and } \sqrt{\frac{3\pi}{2}} \notin A \\
1 - \psi(x) & \text{if } \sqrt{\frac{\pi}{2}} \notin A \text{ and } \sqrt{\frac{3\pi}{2}} \in A \\
1 & \text{if } \left\{\sqrt{\frac{\pi}{2}}, \sqrt{\frac{3\pi}{2}}\right\} \subseteq A
\end{cases}
$$

for every $x \in [0,\infty)$ and $A \in \mathcal{B}([0,\infty))$.

It is easy to see that $P$ is a transition probability. It is also easy to see that $P(x, A) = \psi(x)\delta_{\sqrt{\frac{\pi}{2}}}(A) + (1 - \psi(x))\delta_{\sqrt{\frac{3\pi}{2}}}(A)$ for every $x \in [0,\infty)$ and $A \in \mathcal{B}([0,\infty))$.

Now let $f \in C_b(X)$. Since

$$
\int f(y)P(x,\,\mathrm{d}y) = \int f(y)\,\mathrm{d}\left(\psi(x)\delta_{\sqrt{\frac{\pi}{2}}} + (1 - \psi(x))\delta_{\sqrt{\frac{3\pi}{2}}}\right)(y)
$$
$$
= \psi(x)f\left(\sqrt{\frac{\pi}{2}}\right) + (1 - \psi(x))f\left(\sqrt{\frac{3\pi}{2}}\right),
$$

it follows that (as a function of $x$, of course) $\int f(y)P(x,\,\mathrm{d}y)$ is continuous and bounded. If $T$ is the Markov operator generated by $P$ (using formula (1.1.2)), then Proposition 1.1.4 implies that the map $S : C_b(X) \to C_b(X)$, $Sf(x) = \int f(y)P(x,\,\mathrm{d}y)$ for every $f \in C_b(X)$ and $x \in X$ is a well defined linear operator, and $(S, T)$ is a Markov–Feller pair.

Since $S\psi(x) = \psi(x)\psi\left(\sqrt{\frac{\pi}{2}}\right) + (1 - \psi(x))\psi\left(\sqrt{\frac{3\pi}{2}}\right) = \psi(x)$, it follows that the range of the sequence $(S^n f)_{n \in \mathbb{N}\cup\{0\}}$ is either the set $\{f\}$ (if $Sf = f$), or else the set $\{f, Sf\}$ (if $Sf \neq f$) whenever $f \in C_b(X)$. It follows that $S$ is $C_0(X)$-equicontinuous (actually, $S$ is even $C_b(X)$-equicontinuous).

Now, let $f \in C_0(X)$ be such that $f\left(\sqrt{\frac{\pi}{2}}\right) = 1$ and $f\left(\sqrt{\frac{3\pi}{2}}\right) = 0$ (for example, let $f$ be the function whose graph is the union of the line segments from $(0,0)$ to $(1,0)$; from $(1,0)$ to $\left(\sqrt{\frac{\pi}{2}}, 1\right)$; from $\left(\sqrt{\frac{\pi}{2}}, 1\right)$ to $(2,0)$, and the half-line $\{(x,0)|x \geq 2\}$; clearly, $f \in C_0(X)$ since $f$ is continuous and has compact support). Since $Sf = \psi$ and since $\psi$ is not uniformly continuous, it follows that $(S^n f)_{n \in \mathbb{N}\cup\{0\}}$ is not uniformly equicontinuous (on $X$). Consequently, $S$ is not uniformly $C_0(X)$-equicontinuous; thus, $(S, T)$ is an example of a Markov–Feller pair which is $C_0(X)$-equicontinuous, but fails to be uniformly $C_0(X)$-equicontinuous.

The pair $(S, T)$ that we have discussed here is an example of a Markov–Feller pair defined by an iterated function system (with place-dependent probabilities). The operators $S$ and $T$ that we have constructed here have also been used in our paper [73]. ∎

*Example* 4.1.3. We will now discuss an example of a Markov–Feller pair which is $C_0(X)$-equicontinuous but not $C_b(X)$-equicontinuous.

To this end, let $X$ be the interval $(1, \infty)$ endowed with the distance defined by the absolute value, let $\phi : (1, \infty) \rightarrow (1, \infty)$ be defined by $\phi(x) = 2x$ for every $x \in (1, \infty)$, and let $(S, T)$ be the Markov–Feller pair induced by $\phi$ (for the definition of the Markov–Feller pair induced by a continuous function, see the discussion preceding Example 1.1.9 in Section 1.1). Thus, $S : C_b(X) \rightarrow C_b(X)$ is defined by $Sf(x) = f(2x)$ for every $f \in C_b(X)$ and $x \in (1, \infty)$. In general, $S^n f(x) = f(2^n x)$ for every $f \in C_b(X)$, $n \in \mathbb{N} \cup \{0\}$, and $x \in (1, \infty)$.

We now prove that $S$ is $C_0(X)$-equicontinuous. Since $C_c(X)$ is dense in $C_0(X)$, Lemma 4.1.1 implies that it is enough to prove that $(S^n f)_{n \in \mathbb{N} \cup \{0\}}$ is equicontinuous whenever $f \in C_c(X)$. Thus, let $f \in C_c(X)$, and let $(x_k)_{k \in \mathbb{N}}$ be a sequence of elements of $(1, \infty)$ that converges to some $x \in (1, \infty)$. Then, there exists $r \in \mathbb{R}$, $r > 1$, such that $x \geq r$ and $x_k \geq r$ for every $k \in \mathbb{N}$. Since $f$ has compact support, there exists $M \in \mathbb{R}$, $M > 0$ such that supp $f \subseteq (1, M]$. Clearly, there exists $n_1 \in \mathbb{N}$ such that $2^{n_1} r > M$. Now let $\varepsilon \in \mathbb{R}$, $\varepsilon > 0$. Since $f, Sf, S^2 f, \ldots, S^{n_1 - 1} f$ are continuous functions, there exists $k_\varepsilon \in \mathbb{N}$ such that $|S^n f(x_k) - S^n f(x)| < \varepsilon$ for every $n \in \{0, 1, 2, \ldots, n_1 - 1\}$ and every $k \geq k_\varepsilon$. Since $|S^n f(x_k) - S^n f(x)| = |S^n f(2^n x_k) - S^n f(2^n x)| = 0 < \varepsilon$ for every $n \geq n_1$ and $k \in \mathbb{N}$, it follows that $|S^n f(x_k) - S^n f(x)| < \varepsilon$ for every $n \in \mathbb{N} \cup \{0\}$ and $k \geq k_\varepsilon$. Thus, $(S^n f)_{n \in \mathbb{N} \cup \{0\}}$ is equicontinuous for every $f \in C_c(X)$.

In order to prove that $S$ is not $C_b(X)$-equicontinuous, let $g \in C_b(X)$, $g(x) = \sin x$ for all $x \in X$. We will prove that $(S^n g)_{n \in \mathbb{N} \cup \{0\}}$ is not equicontinuous. To this end, consider the sequence $(x_k)_{k \in \mathbb{N}}$, $x_k = \dfrac{\pi}{2} + \dfrac{1}{k}$ for every $k \in \mathbb{N}$, and let $\varepsilon_0 = \dfrac{\sin 1}{2}$. Since $\lim\limits_{k \to \infty} x_k = \dfrac{\pi}{2}$, and since

$$\left| S^n g\left(x_{2^n}\right) - S^n g\left(\frac{\pi}{2}\right) \right| = \sin 1 > \varepsilon_0$$

for every $n \in \mathbb{N}$, it follows that $(S^n g)_{n \in \mathbb{N} \cup \{0\}}$ is not equicontinuous. ∎

Even though the notion of $C_0(X)$-equicontinuity is the most general notion of equicontinuity mentioned in this work, it turns out that all the results that involve the equicontinuity of Markov–Feller pairs in this volume can be stated in terms of $C_0(X)$-equicontinuity. Thus, from now on we will deal only with $C_0(X)$-equicontinuity, and we will often refer to a $C_0(X)$-equicontinuous Markov–Feller pair $(S, T)$ (or to the operator $S$), simply as an equicontinuous pair (or operator).

The class of $C_0(X)$-equicontinuous Markov–Feller pairs is fairly large. If $(X, d)$ is discrete (that is, if the set $\{x\}$ is open for every $x \in X$), then every Markov–Feller pair defined on $(X, d)$ is even $C_b(X)$-equicontinuous. Thus, the Markov–Feller pairs described in Example 1.1.9, Example 1.1.10, and Example 1.1.14 are all $C_b(X)$-equicontinuous. Naturally, many $C_0(X)$-equicontinuous Markov–Feller pairs are defined on spaces that are not discrete; some examples are the Markov–Feller pairs of Example 1.1.11 (the Markov–Feller pairs induced by rotations of the unit circle), Example 4.1.2, and Example 4.1.3. Of course, not every Markov–Feller pair is $C_0(X)$-equicontinuous; except for the trivial case when $\Lambda$ is a singleton (has only one element), the Markov–Feller pairs induced by symbolic flows (see Example 1.1.13) fail to be $C_0(X)$-equicontinuous.

For future reference we will now discuss briefly several simple general facts related to $C_0(X)$-equicontinuity.

**Proposition 4.1.4.** *As usual, we assume given the Markov–Feller pair $(S,T)$ (not necessarily $C_0(X)$-equicontinuous). Let $f \in C_b(X)$. Then the following assertions are equivalent:*

(a) *The sequence $(S^n f)_{n \in \mathbb{N} \cup \{0\}}$ converges uniformly on the compact subsets of $X$.*

(b) *The sequence $(S^n f)_{n \in \mathbb{N} \cup \{0\}}$ is uniformly Cauchy on the compact subsets of $X$.*

*Proof.* The proof consists on a straightforward application of Proposition 1.3.8. □

**Proposition 4.1.5.** *Assume that $f \in C_b(X)$ satisfies (a) (or (b)) of Proposition 4.1.4, and let $\overline{f} : X \to \mathbb{R}$ be the uniform limit of $(S^n f)_{n \in \mathbb{N} \cup \{0\}}$ on the compact subsets of $X$. Then $\overline{f} \in C_b(X)$. Also, the sequence $(S^n f)_{n \in \mathbb{N} \cup \{0\}}$ is equicontinuous.*

*Proof.* Since $S$ is a (positive) contraction of $C_b(X)$, it follows that $(S^n f)_{n \in \mathbb{N} \cup \{0\}}$ is a bounded sequence of elements of $C_b(X)$; hence, we can apply Proposition 1.3.9. Using (a) and (b) of Proposition 1.3.9, we obtain that $\overline{f} \in C_b(X)$ and that $(S^n f)_{n \in \mathbb{N} \cup \{0\}}$ is equicontinuous, respectively. □

The above two propositions have the following obvious consequence:

**Corollary 4.1.6.** *If the Markov–Feller pair $(S,T)$ has the property that condition (a) (or (b)) of Proposition 4.1.4 is satisfied by every $f \in C_0(X)$, then $(S,T)$ is $C_0(X)$-equicontinuous.*

Now, our goal is to prove that if $(S,T)$ is $C_0(X)$-equicontinuous and has invariant probabilities, then the nonemptiness of any of the sets $\gamma_{cp}$, $\gamma_c$, $\gamma_0$, or $\gamma$ implies the unique ergodicity of $(S,T)$. To this end, we need some preparation.

We say that $(S,T)$ (or $T$) has the *e.m.d.s.* property (has the property that its **e**rgodic **m**easures are **d**isjointly **s**upported) if $(\operatorname{supp}\mu) \cap (\operatorname{supp}\nu) = \emptyset$ whenever $\mu$ and $\nu$ are distinct $T$-invariant ergodic probability measures. Note that if $T$ does not have invariant probabilities (hence, $T$ does not have invariant ergodic probabilities), or if $T$ is uniquely ergodic (so, $T$ has exactly one ergodic measure (see Theorem 1.2.2 or Corollary 3.3.3)), then $T$ has the e.m.d.s. property.

**Theorem 4.1.7.** *If the Markov–Feller pair $(S, T)$ is $C_0(X)$-equicontinuous, then $(S, T)$ has the e.m.d.s. property.*

*Proof.* In view of our remarks on the e.m.d.s. property, it is enough to prove the theorem under the assumption that $T$ has at least two distinct ergodic measures.

Let $\mu$ and $\nu$ be two distinct $T$-invariant ergodic measures, assume that $(\operatorname{supp} \mu) \cap (\operatorname{supp} \nu) \neq \emptyset$, and let $z \in (\operatorname{supp} \mu) \cap (\operatorname{supp} \nu)$. Since $\mu$ and $\nu$ are ergodic, Theorem 3.3.5 implies that there exist $x \in \Gamma_{cp}$ and $y \in \Gamma_{cp}$ such that $\mu = \varepsilon_x$, $\nu = \varepsilon_y$, $\varepsilon_x([x]) = 1$, and $\varepsilon_y([y]) = 1$. Therefore, $\operatorname{supp} \mu \subseteq \overline{[x]}$ and $\operatorname{supp} \nu \subseteq \overline{[y]}$. Thus, $z \in \overline{[x]} \cap \overline{[y]}$; accordingly, there exist two convergent sequences $(x_k)_{k \in \mathbb{N}}$ and $(y_k)_{k \in \mathbb{N}}$ of elements of $[x]$ and $[y]$, respectively, such that $\lim_{k \to \infty} x_k = \lim_{k \to \infty} y_k = z$.

Since $\mu \neq \nu$, and since both $\mu$ and $\nu$ belong to the topological dual $\mathcal{M}(X)$ of $C_0(X)$, there exists $f \in C_0(X)$ such that $\langle f, \mu \rangle \neq \langle f, \nu \rangle$. Set $\varepsilon_0 = \dfrac{|\langle f, \mu \rangle - \langle f, \nu \rangle|}{8}$.

Since $S$ is $C_0(X)$-equicontinuous, there exist $k_{\varepsilon_0}^{(1)} \in \mathbb{N}$ and $k_{\varepsilon_0}^{(2)} \in \mathbb{N}$ such that for every $n \in \mathbb{N} \cup \{0\}$ we have $|S^n f(x_k) - S^n f(z)| < \varepsilon_0$ for every $k \geq k_{\varepsilon_0}^{(1)}$, and $|S^n f(y_k) - S^n f(z)| < \varepsilon_0$ for every $k \geq k_{\varepsilon_0}^{(2)}$.

Set $k_0 = \max\{k_{\varepsilon_0}^{(1)}, k_{\varepsilon_0}^{(2)}\}$.

Since $x_{k_0} \in [x]$ and $y_{k_0} \in [y]$, it follows that $\displaystyle\lim_{m \to \infty} \frac{1}{m} \sum_{n=0}^{m-1} S^n f(x_{k_0}) = \langle f, \mu \rangle$

and $\displaystyle\lim_{m \to \infty} \frac{1}{m} \sum_{n=0}^{m-1} S^n f(y_{k_0}) = \langle f, \nu \rangle$. Thus, there exists $m_0 \in \mathbb{N}$ large enough such

that $\left| \dfrac{1}{m} \displaystyle\sum_{n=0}^{m-1} S^n f(x_{k_0}) - \langle f, \mu \rangle \right| < \varepsilon_0$ and $\left| \dfrac{1}{m} \displaystyle\sum_{n=0}^{m-1} S^n f(y_{k_0}) - \langle f, \nu \rangle \right| < \varepsilon_0$ for every $m \geq m_0$.

We obtain that

$$|\langle f, \mu \rangle - \langle f, \nu \rangle| \leq \left| \langle f, \mu \rangle - \frac{1}{m_0} \sum_{n=0}^{m_0-1} S^n f(x_{k_0}) \right|$$

$$+ \left| \frac{1}{m_0} \sum_{n=0}^{m_0-1} S^n f(x_{k_0}) - \frac{1}{m_0} \sum_{n=0}^{m_0-1} S^n f(z) \right|$$

$$+ \left| \frac{1}{m_0} \sum_{n=0}^{m_0-1} S^n f(z) - \frac{1}{m_0} \sum_{n=0}^{m_0-1} S^n f(y_{k_0}) \right|$$

$$+ \left| \frac{1}{m_0} \sum_{n=0}^{m_0-1} S^n f(y_{k_0}) - \langle f, \nu \rangle \right| < \varepsilon_0 + \frac{1}{m_0} \sum_{n=0}^{m_0-1} |S^n f(x_{k_0}) - S^n f(z)|$$

$$+ \frac{1}{m_0} \sum_{n=0}^{m_0-1} |S^n f(z) - S^n f(y_{k_0})| + \varepsilon_0 < \varepsilon_0 + \varepsilon_0 + \varepsilon_0 + \varepsilon_0 = 4\varepsilon_0.$$

We obtained a contradiction since $|\langle f, \mu \rangle - \langle f, \nu \rangle| = 8\varepsilon_0$ and $\varepsilon_0 > 0$. The contradiction stems from our assumption that $(\text{supp } \mu) \cap (\text{supp } \nu) \neq \emptyset$. Accordingly, every two distinct $T$-invariant ergodic measures have disjoint supports; that is, $T$ has the e.m.d.s. property.                                                                    □

We are now in the position to prove the criterion for unique ergodicity of $C_0(X)$-equicontinuous Markov–Feller pairs that we have been aiming for.

**Theorem 4.1.8.** *Assume that $(S,T)$ has invariant probabilities, and is a $C_0(X)$-equicontinuous Markov–Feller pair. Then $(S,T)$ is uniquely ergodic if and only if at least one of the sets $\gamma_{cp}$, $\gamma_c$, $\gamma_0$, or $\gamma$ is nonempty. If any of these sets is nonempty, then all of them are nonempty and equal to the support of the unique $T$-invariant probability.*

*Proof.* If $(S,T)$ is uniquely ergodic, then, by Theorem 3.1.1, each of the sets $\gamma_{cp}$, $\gamma_c$, $\gamma_0$, and $\gamma$ is nonempty and equal to the support of the unique $T$-invariant probability.

We now prove that if at least one of the four sets $\gamma_{cp}$, $\gamma_c$, $\gamma_0$, or $\gamma$ is nonempty, then $(S,T)$ is uniquely ergodic. To this end, we note (as pointed out in the proof of Theorem 3.1.1) that $\gamma_{cp} \supseteq \gamma_c \supseteq \gamma_0 \supseteq \gamma$; hence, if one of the four sets is nonempty, then $\gamma_{cp}$ is nonempty. Therefore, it is enough to prove that if $\gamma_{cp}$ is nonempty, then $(S,T)$ is uniquely ergodic.

Thus, assume that $\gamma_{cp}$ is nonempty, and assume also that $(S,T)$ is not uniquely ergodic. Since $(S,T)$ has invariant probabilities, Theorem 1.2.2 implies that there exist at least two distinct $T$-invariant ergodic measures. By Theorem 3.3.5 there exist $x \in \Gamma_{cp}$, $y \in \Gamma_{cp}$, $x \neq y$ such that $\varepsilon_x$ and $\varepsilon_y$ are distinct $T$-invariant ergodic measures. Using the comment made in the last paragraph of Section 3.3, we obtain that $[x] \subseteq \Gamma_1 \subseteq \Gamma_{cp}$ and $[y] \subseteq \Gamma_1 \subseteq \Gamma_{cp}$. Thus, by Theorem 2.2.2, $\text{supp } \varepsilon_x \supseteq \gamma_{cp}$ and $\text{supp } \varepsilon_y \supseteq \gamma_{cp}$. Since $(S,T)$ is $C_0(X)$-equicontinuous, Theorem 4.1.7 implies that $\text{supp } \varepsilon_x$ and $\text{supp } \varepsilon_y$ are disjoint. Since $\gamma_{cp} \neq \emptyset$, we obtained a contradiction which stems from our assumption that $(S,T)$ is not uniquely ergodic.

It is now obvious that if any of the four sets $\gamma_{cp}$, $\gamma_c$, $\gamma_0$, and $\gamma$ is nonempty, then each of the four sets is nonempty and equal to the support of the unique invariant probability of $(S,T)$.                                                                    □

Let us illustrate the use of Theorem 4.1.8 in the case of irrational rotations of the unit circle. Let $a \in \mathbb{R}/\mathbb{Z}$, assume that the equivalence class $a$ contains irrational numbers, and let $(S_a, T_a)$ be the Markov–Feller pair induced by the rotation of the unit circle $\mathbb{R}/\mathbb{Z}$ by $a$ (see Example 1.1.11). It is well known that $(S_a, T_a)$ is uniquely ergodic (see, for example, p. 178 of Krengel [32]). If we assume known that the orbit $\mathcal{O}(x)$ of $x$ (under the action of $T_a$) is dense in $\mathbb{R}/\mathbb{Z}$ whenever $x \in \mathbb{R}/\mathbb{Z}$ (for a proof that these orbits are dense in $\mathbb{R}/\mathbb{Z}$ see, for example, pp. 12–13 of Krengel [32], or Theorem 8.3, p. 53, Section 2.8 of Robinson [58]), then we can use Theorem 4.1.8 to prove the unique ergodicity of $(S_a, T_a)$. Indeed, it is easy to see that the Lebesgue (Haar) measure on $\mathbb{R}/\mathbb{Z}$ is an invariant probability for $(S_a, T_a)$, so $(S_a, T_a)$ has invariant probabilities. Since $(S_a, T_a)$ is obviously

$C_0(\mathbb{R}/\mathbb{Z})$-equicontinuous, Theorem 4.1.8 implies that $(S_a, T_a)$ is uniquely ergodic.

Theorem 4.1.8 can be reformulated in terms of the universal elements that were defined in Section 3.1. To this end, given (as usual in this section) a Markov–Feller pair $(S, T)$, let $\mathcal{U}_{cp}$, $\mathcal{U}_c$, $\mathcal{U}_0$, and $\mathcal{U}$ be the sets of all universal elements with respect to $\Gamma_{cp}$, $\Gamma_c$, $\Gamma_0$, and $\Gamma$, respectively, generated by $(S, T)$. In terms of $\mathcal{U}_{cp}$, $\mathcal{U}_c$, $\mathcal{U}_0$, and $\mathcal{U}$, Theorem 4.1.8 becomes:

**Theorem 4.1.9.** *Assume that $(S, T)$ has invariant probabilities, and is $C_0(X)$-equicontinuous. Then $(S, T)$ is uniquely ergodic if and only if at least one of the sets $\mathcal{U}_{cp}$, $\mathcal{U}_c$, $\mathcal{U}_0$, or $\mathcal{U}$ is nonempty. If any of these sets is nonempty, then all of them are nonempty, and equal to the support of the unique $T$-invariant probability.*

*Proof.* The proof is obvious, since $\mathcal{U}_{cp} = \gamma_{cp}$, $\mathcal{U}_c = \gamma_c$, $\mathcal{U}_0 = \gamma_0$, and $\mathcal{U} = \gamma$. $\square$

Theorem 4.1.9 has the following consequences (recall that we say that an element $x \in X$ is universal if it is universal with respect to $X$):

**Corollary 4.1.10.** *Assume that the Markov–Feller pair $(S, T)$ defined on $(X, d)$ is $C_0(X)$-equicontinuous.*

(a) *If $(S, T)$ has invariant probabilities and universal elements, then $(S, T)$ is uniquely ergodic. In this case, the set of all universal elements is included in the support of the unique $T$-invariant probability.*

(b) *If $X$ is compact, then $(S, T)$ is uniquely ergodic if and only if $(S, T)$ has universal elements. In this case, the support of the unique $T$-invariant probability is the set of all universal elements of $(S, T)$.*

*Proof.* (a) If $(S, T)$ has universal elements, then these elements are universal with respect to each of the sets $\Gamma_{cp}$, $\Gamma_c$, $\Gamma_0$, and $\Gamma$. Thus, Theorem 4.1.9 implies that $(S, T)$ is uniquely ergodic, and the universal elements (with respect to $X$) of $(S, T)$ belong to the support of the unique $T$-invariant probability.

(b) If $X$ is compact, then $\Gamma = X$, so $\gamma = \mathcal{U}$. By Theorem 4.1.9, the Markov–Feller pair $(S, T)$ is uniquely ergodic if and only if $\mathcal{U} \neq \emptyset$; in this case, the support of the unique $T$-invariant probability is equal to $\mathcal{U}$. $\square$

We will now discuss another criterion for the unique ergodicity of $C_0(X)$-equicontinuous Markov–Feller pairs. In order to discuss the criterion, recall (see Section 2.2) that given a Markov–Feller pair $(S, T)$ defined on $(X, d)$, and given $x \in X$, the orbit-closure of $x$ under the action of $T$ is denoted by $\overline{\mathcal{O}(x)}$ and is defined by $\overline{\mathcal{O}(x)} = \bigcup_{n=0}^{\infty} \mathrm{supp}\,(T^n \delta_x)$.

**Theorem 4.1.11.** *Assume that the Markov–Feller pair $(S, T)$ is $C_0(X)$-equicontinuous and has invariant probabilities. If $\overline{\mathcal{O}(x)} \cap \overline{\mathcal{O}(y)} \neq \emptyset$ for every $x \in X$ and $y \in X$, then $(S, T)$ is uniquely ergodic.*

*Proof.* Assume that $(S, T)$ is not uniquely ergodic. Then, as pointed out at the beginning of the proof of Theorem 4.1.8 there exist two distinct $T$-invariant ergodic measures, say $\mu$ and $\nu$ (the existence of $\mu$ and $\nu$ is assured by the fact that $(S, T)$ has invariant probabilities and is not uniquely ergodic).

Now let $x \in \operatorname{supp} \mu$ and $y \in \operatorname{supp} \nu$. By Proposition 1.1.7, $\operatorname{supp}(T^n \delta_x) \subseteq \operatorname{supp}(T^n \mu) = \operatorname{supp} \mu$ and $\operatorname{supp}(T^n \delta_y) \subseteq \operatorname{supp}(T^n \nu) = \operatorname{supp} \nu$ for every $n \in \mathbb{N}$. Consequently, $\mathcal{O}(x) \subseteq \operatorname{supp} \mu$ and $\mathcal{O}(y) \subseteq \operatorname{supp} \nu$. Since the support of a measure is a closed set, it follows that $\overline{\mathcal{O}(x)} \subseteq \operatorname{supp} \mu$ and $\overline{\mathcal{O}(y)} \subseteq \operatorname{supp} \nu$. Taking into consideration that (by Theorem 4.1.7) $(S, T)$ has the e.m.d.s. property, we obtain that $(\operatorname{supp} \mu) \cap (\operatorname{supp} \nu) = \emptyset$; hence, $\overline{\mathcal{O}(x)} \cap \overline{\mathcal{O}(y)} = \emptyset$. We have obtained a contradiction, which stems from our assumption that $(S, T)$ is not uniquely ergodic. $\qquad \square$

In Theorem 4.1.11, the assumption that $(S, T)$ has invariant probabilities cannot be omitted; that is, if $(S, T)$ is $C_0(X)$-equicontinuous and has the property that $\overline{\mathcal{O}(x)} \cap \overline{\mathcal{O}(y)} \neq \emptyset$ for every $x \in X$ and $y \in X$, it may happen that $(S, T)$ is not uniquely ergodic, simply because it may also happen that $(S, T)$ does not have invariant probabilities. A case in point is the Markov–Feller pair of Example 1.1.10 (in Section 1.1).

## 4.2   A Diagonalization Procedure: Technical Preliminaries for Mean Ergodic Theorems

In the next section (Section 4.3) we will prove two mean ergodic theorems for $C_0(X)$-equicontinuous Markov–Feller pairs: a weak* mean ergodic theorem and a pointwise mean ergodic theorem. Roughly speaking, the weak* mean ergodic theorem is proved by a somewhat involved diagonalization procedure (while the other ergodic theorem is just a simple (but interesting) consequence of the weak* ergodic theorem). Our goal in this section is to discuss several facts that deal with the diagonalization process, and that will be used in the proof of the weak* mean ergodic theorem.

As usual, throughout this section, we assume given a locally compact separable metric space $(X, d)$ and a Markov–Feller pair $(S, T)$ defined on $(X, d)$.

Recall that given a sequence $(\mu_n)_{n \in \mathbb{N}}$ of elements of $\mathcal{M}(X)$ and $\mu \in \mathcal{M}(X)$, we say that $(\mu_n)_{n \in \mathbb{N}}$ converges to $\mu$ in the weak* topology of $\mathcal{M}(X)$ if the sequence (of real numbers) $(\langle f, \mu_n \rangle)_{n \in \mathbb{N}}$ converges to $\langle f, \mu \rangle$ whenever $f \in C_0(X)$.

**Proposition 4.2.1.** *Let $\mu^* \in \mathcal{M}(X)$, let $(\mu_n)_{n \in \mathbb{N}}$ be a sequence of elements of $\mathcal{M}(X)$, and assume that for every subsequence $(\mu_{n_k})_{k \in \mathbb{N}}$ of $(\mu_n)_{n \in \mathbb{N}}$ there exists a subsequence $\left(\mu_{n_{k_l}}\right)_{l \in \mathbb{N}}$ of $(\mu_{n_k})_{k \in \mathbb{N}}$ such that $\left(\mu_{n_{k_l}}\right)_{l \in \mathbb{N}}$ converges weak\* (that is, in the weak\* topology of $\mathcal{M}(X)$) to $\mu^*$. Then $(\mu_n)_{n \in \mathbb{N}}$ converges weak\* to $\mu^*$.*

*Proof.* Assume that $(\mu_n)_{n \in \mathbb{N}}$ does not converge in the weak* topology of $\mathcal{M}(X)$ to $\mu^*$. Then there exists $f \in C_0(X)$ such that the sequence $(\langle f, \mu_n \rangle)_{n \in \mathbb{N}}$ does not converge to $\langle f, \mu^* \rangle$. Therefore, there exist $\varepsilon_0 \in \mathbb{R}$, $\varepsilon_0 > 0$, and a subsequence $(\mu_{n_k})_{k \in \mathbb{N}}$ of $(\mu_n)_{n \in \mathbb{N}}$ such that $|\langle f, \mu_{n_k} \rangle - \langle f, \mu^* \rangle| \geq \varepsilon_0$ for every $k \in \mathbb{N}$. But in this case it is easy to see that no subsequence $\left( \mu_{n_{k_l}} \right)_{l \in \mathbb{N}}$ of $(\mu_{n_k})_{k \in \mathbb{N}}$ converges weak* to $\mu^*$. We have obtained a contradiction which stems from our assumption that $(\mu_n)_{n \in \mathbb{N}}$ does not converge weak* to $\mu^*$. $\qquad \square$

Note that the arguments offered in the proof of the above lemma can be used to prove a slightly more general assertion than the one stated in the lemma. More precisely, it is easy to see that if $\mu^* \in \mathcal{M}(X)$, and if $(\mu_n)_{n \in \mathbb{N}}$ is a sequence of elements of $\mathcal{M}(X)$ such that for every subsequence $(\mu_{n_k})_{k \in \mathbb{N}}$ of $(\mu_n)_{n \in \mathbb{N}}$ and for every $f \in C_0(X)$ there exists a subsequence $\left( \mu_{n_{k_l}} \right)_{l \in \mathbb{N}}$ of $(\mu_{n_k})_{k \in \mathbb{N}}$ such that $\left( \left\langle f, \mu_{n_{k_l}} \right\rangle \right)_{l \in \mathbb{N}}$ converges to $\langle f, \mu^* \rangle$, then $(\mu_n)_{n \in \mathbb{N}}$ converges weak* to $\mu^*$. However, we will not need this more general assertion in the future.

Recall (see the subsection *Almost Everywhere Convergence Results* of Section 1.2) that we use the notation $A_n(Q) = \dfrac{1}{n} \displaystyle\sum_{k=0}^{n-1} Q^k$ whenever $E$ is a Banach space, $Q : E \to E$ is a linear operator, and $n \in \mathbb{N}$.

**Proposition 4.2.2.** *Assume that the Markov–Feller pair $(S,T)$ is $C_0(X)$-equicontinuous, and let $(A_{n_k}(S))_{k \in \mathbb{N}}$ be a subsequence of $(A_n(S))_{n \in \mathbb{N}}$. Then there exists a subsequence $\left( A_{n_{k(l)}}(S) \right)_{l \in \mathbb{N}}$ of $(A_{n_k}(S))_{k \in \mathbb{N}}$ such that for every $f \in C_0(X)$ and $x \in X$, the sequence $\left( A_{n_{k(l)}}(S)f(x) \right)_{l \in \mathbb{N}}$ converges.*

*Proof.* Since $X$ is separable, and since (by Theorem 1.3.3) the space $C_0(X)$ is also separable, there exist two countable subsets $D$ and $\mathcal{D}$ of $X$ and $C_0(X)$, respectively, such that $D$ is dense in $X$ and $\mathcal{D}$ is dense in $C_0(X)$.

Since $\mathcal{D} \times D$ is a countable set, we can construct a sequence $(g_m, y_m)_{m \in \mathbb{N}}$ such that $g_m \in \mathcal{D}$ and $y_m \in D$ for every $m \in \mathbb{N}$, and such that the range of $(g_m, y_m)_{m \in \mathbb{N}}$ is the entire set $\mathcal{D} \times D$.

Since $S$ is a contraction of $C_b(X)$, and $g_1$ is a bounded function, it follows that $(A_{n_k}(S)g_1(y_1))_{k \in \mathbb{N}}$ is a bounded sequence of real numbers. Therefore, there exists a convergent subsequence $\left( A_{n_{k_{j^{(1)}}}}(S)g_1(y_1) \right)_{j^{(1)} \in \mathbb{N}}$ of $(A_{n_k}(S)g_1(y_1))_{k \in \mathbb{N}}$.

Thus, there exists $j_{[1]}^{(1)} \in \mathbb{N}$ such that $\left| A_{n_{k_{j^{(1)}}}}(S)g_1(y_1) - A_{n_{k_{j^{(1)'}}}}(S)g_1(y_1) \right| < 1$ for every $j^{(1)} \geq j_{[1]}^{(1)}$ and $j^{(1)'} \geq j_{[1]}^{(1)}$. Set $n_{k_{(1)}} = n_{k_{j_{[1]}^{(1)}}}$.

Since $S$ is a contraction of $C_b(X)$ and $g_2$ is a bounded function, it follows that $\left( A_{n_{k_{j^{(1)}}}}(S)g_2(y_2) \right)_{j^{(1)} \geq j_{[1]}^{(1)}}$ is a bounded sequence, so there exists a convergent subsequence $\left( A_{n_{k_{j^{(1)}_{j^{(2)}}}}} (S)g_2(y_2) \right)_{j^{(2)} \in \mathbb{N}}$ of $\left( A_{n_{k_{j^{(1)}}}} (S)g_2(y_2) \right)_{j^{(1)} \geq j^{(1)}_{[1]}}$. Therefore,

there exists $j^{(2)}_{[2]} \in \mathbb{N}$ such that

$$\left| A_{n_{k_{j^{(1)}_{j^{(2)}}}}} (S)g_i(y_i) - A_{n_{k_{j^{(1)}_{j^{(2)'}}}}} (S)g_i(y_i) \right| < \frac{1}{2}$$

for every $j^{(2)} \geq j^{(2)}_{[2]}$, $j^{(2)'} \geq j^{(2)}_{[2]}$, and $i = 1, 2$. Set $n_{k_{(2)}} = n_{k_{j^{(1)}_{j^{(2)}_{[2]}}}}$.

Now, in general, assume that we have constructed $n_{k_{(1)}}, n_{k_{(2)}}, n_{k_{(3)}}, \ldots, n_{k_{(l)}}$. Since $S$ is a contraction of $C_b(X)$ and $g_{l+1}$ is a bounded function, it follows that

$$\left( A_{n_{k_{j^{(1)}_{j^{(2)}_{j^{(3)}_{\cdots_{j^{(l)}}}}}}}} (S)g_{l+1}(y_{l+1}) \right)_{j^{(l)} \geq j^{(l)}_{[l]}} \quad \text{is a bounded sequence, so there exists a con-}$$

vergent subsequence

$$\left( A_{n_{k_{j^{(1)}_{j^{(2)}_{j^{(3)}_{\cdots_{j^{(l)}_{j^{(l+1)}}}}}}}}} (S)g_{l+1}(y_{l+1}) \right)_{j^{(l+1)} \in \mathbb{N}} \quad \text{of} \quad \left( A_{n_{k_{j^{(1)}_{j^{(2)}_{j^{(3)}_{\cdots_{j^{(l)}}}}}}}} (S)g_{l+1}(y_{l+1}) \right)_{j^{(l)} \geq j^{(l)}_{[l]}} .$$

Therefore, there exists $j^{(l+1)}_{[l+1]} \in \mathbb{N}$ such that

$$\left| A_{n_{k_{j^{(1)}_{j^{(2)}_{j^{(3)}_{\cdots_{j^{(l)}_{j^{(l+1)}}}}}}}}} (S)g_i(y_i) - A_{n_{k_{j^{(1)}_{j^{(2)}_{j^{(3)}_{\cdots_{j^{(l)}_{j^{(l+1)'}}}}}}}}} (S)g_i(y_i) \right| < \frac{1}{l+1}$$

for every $j^{(l+1)} \geq j^{(l+1)}_{[l+1]}$, $j^{(l+1)'} \geq j^{(l+1)}_{[l+1]}$, and $i = 1, 2, 3, \ldots, l+1$. Set $n_{k_{(l+1)}} = n_{j^{(1)}_{j^{(2)}_{j^{(3)}_{\cdots_{j^{(l)}_{j^{(l+1)}_{[l+1]}}}}}}}$.

It is easy to see that the sequence $\left( A_{n_{k_{(l)}}} (S)g(y) \right)_{l \in \mathbb{N}}$ converges whenever $(g, y) \in \mathcal{D} \times \mathcal{D}$. Indeed, if $(g, y) \in \mathcal{D} \times \mathcal{D}$, then the sequence $\left( A_{n_{k_{(l)}}} (S)g(y) \right)_{l \in \mathbb{N}}$ is

Cauchy (hence, convergent) since there exists $m \in \mathbb{N}$ such that $(g, y) = (g_m, y_m)$ (where $(g_i, y_i)_{i \in \mathbb{N}}$ is the sequence used to construct the $n_{k_{(l)}}$'s), and since for every $\varepsilon \in \mathbb{R}$, $\varepsilon > 0$ there exists $l_\varepsilon \in \mathbb{N}$, $l_\varepsilon \geq m$ such that $\dfrac{1}{l_\varepsilon} < \varepsilon$. Therefore, in view of the construction of the sequence $(n_{k_{(l)}})_{l \in \mathbb{N}}$, it follows that

$$\left| A_{n_{k_{(l)}}}(S)g_m(y_m) - A_{n_{k_{(l')}}}(S)g_m(y_m) \right| < \frac{1}{l_\varepsilon} < \varepsilon$$

for every $l \geq l_\varepsilon$ and $l' \geq l_\varepsilon$.

We now prove that the sequence $\left( A_{n_{k_{(l)}}}(S)g(x) \right)_{l \in \mathbb{N}}$ is Cauchy (convergent) whenever $g \in \mathcal{D}$ and $x \in X$. To this end, let $g \in \mathcal{D}$, $x \in X$, and $\varepsilon \in \mathbb{R}$, $\varepsilon > 0$. Since $\mathcal{D}$ is dense in $X$, there exists a sequence $(y_i)_{i \in \mathbb{N}}$ of elements of $\mathcal{D}$ that converges to $x$. Since $S$ is $C_0(X)$-equicontinuous, there exists $i_\varepsilon \in \mathbb{N}$ such that $|S^m g(y_i) - S^m g(x)| < \dfrac{\varepsilon}{3}$ for every $i \geq i_\varepsilon$ and $m \in \mathbb{N} \cup \{0\}$. Taking into consideration that the sequence $\left( A_{n_{k_{(l)}}}(S)g(y_{i_\varepsilon}) \right)_{l \in \mathbb{N}}$ converges (because $(g, y_{i_\varepsilon}) \in \mathcal{D} \times \mathcal{D}$), we obtain that there exists $l_\varepsilon \in \mathbb{N}$ such that $\left| A_{n_{k_{(l)}}}(S)g(y_{i_\varepsilon}) - A_{n_{k_{(l')}}}(S)g(y_{i_\varepsilon}) \right| < \dfrac{\varepsilon}{3}$ for every $l \geq l_\varepsilon$ and $l' \geq l_\varepsilon$. It follows that

$$\left| A_{n_{k_{(l)}}}(S)g(x) - A_{n_{k_{(l')}}}(S)g(x) \right| \leq \left| A_{n_{k_{(l)}}}(S)g(x) - A_{n_{k_{(l)}}}(S)g(y_{i_\varepsilon}) \right|$$

$$+ \left| A_{n_{k_{(l)}}}(S)g(y_{i_\varepsilon}) - A_{n_{k_{(l')}}}(S)g(y_{i_\varepsilon}) \right| + \left| A_{n_{k_{(l')}}}(S)g(y_{i_\varepsilon}) - A_{n_{k_{(l')}}}(S)g(x) \right|$$

$$\leq \frac{1}{n_{k_{(l)}}} \sum_{m=0}^{n_{k_{(l)}}-1} |S^m g(x) - S^m g(y_{i_\varepsilon})| + \left| A_{n_{k_{(l)}}}(S)g(y_{i_\varepsilon}) - A_{n_{k_{(l')}}}(S)g(y_{i_\varepsilon}) \right|$$

$$+ \frac{1}{n_{k_{(l')}}} \sum_{m=0}^{n_{k_{(l')}}-1} |S^m g(y_{i_\varepsilon}) - S^m g(x)| < \frac{\varepsilon}{3} + \frac{\varepsilon}{3} + \frac{\varepsilon}{3} = \varepsilon$$

for every $l \geq l_\varepsilon$ and $l' \geq l_\varepsilon$. Thus, the sequence $\left( A_{n_{k_{(l)}}}(S)g(x) \right)_{l \in \mathbb{N}}$ is convergent whenever $g \in \mathcal{D}$ and $x \in X$.

In order to complete the proof of the proposition, we only have to show that $\left( A_{n_{k_{(l)}}}(S)f(x) \right)_{l \in \mathbb{N}}$ is a Cauchy (convergent) sequence whenever $f \in C_0(X)$ and $x \in X$. So, let $f \in C_0(X)$, $x \in X$, and $\varepsilon \in \mathbb{R}$, $\varepsilon > 0$. Since $\mathcal{D}$ is dense in $C_0(X)$, there exists $g \in \mathcal{D}$ such that $\|f - g\| < \dfrac{\varepsilon}{3}$. Since the sequence $\left( A_{n_{k_{(l)}}}(S)g(x) \right)_{l \in \mathbb{N}}$ converges, there exists $l_\varepsilon \in \mathbb{N}$ such that $\left| A_{n_{k_{(l)}}}(S)g(x) - A_{n_{k_{(l')}}}(S)g(x) \right| < \dfrac{\varepsilon}{3}$ for every $l \geq l_\varepsilon$ and $l' \geq l_\varepsilon$. Taking into consideration that $\|S^m f - S^m g\| < \dfrac{\varepsilon}{3}$ for every $m \in \mathbb{N} \cup \{0\}$ (since $S$ is a contraction), we obtain that

$$\left| A_{n_{k_{(l)}}}(S)f(x) - A_{n_{k_{(l')}}}(S)f(x) \right| \leq \left| A_{n_{k_{(l)}}}(S)f(x) - A_{n_{k_{(l)}}}(S)g(x) \right|$$

$$+ \left| A_{n_{k_{(l)}}}(S)g(x) - A_{n_{k_{(l')}}}(S)g(x) \right| + \left| A_{n_{k_{(l')}}}(S)g(x) - A_{n_{k_{(l')}}}(S)f(x) \right|$$

$$\leq \frac{1}{n_{k_{(l)}}} \sum_{m=0}^{n_{k_{(l)}}-1} \|S^m(f-g)(x)\| + \left| A_{n_{k_{(l)}}}(S)g(x) - A_{n_{k_{(l')}}}(S)g(x) \right|$$

$$+ \frac{1}{n_{k_{(l')}}} \sum_{m=0}^{n_{k_{(l')}}-1} \|S^m(f-g)(x)\| < \frac{\varepsilon}{3} + \frac{\varepsilon}{3} + \frac{\varepsilon}{3} = \varepsilon$$

for every $l \geq l_\varepsilon$ and $l' \geq l_\varepsilon$. Thus, the sequence $\left( A_{n_{k_{(l)}}}(S)f(x) \right)_{l \in \mathbb{N}}$ is convergent. $\qquad \square$

Let $\mathbb{N}^{\mathbb{N}}$ be the collection of all sequences of natural numbers, and set

$$\mathcal{A} = \left\{ \alpha \in \mathbb{N}^{\mathbb{N}} \;\middle|\; \begin{array}{l} \alpha = (n_i)_{i \in \mathbb{N}} \text{ is a strictly increasing sequence of natural} \\ \text{numbers such that the sequence } (A_{n_i}(S)f(x))_{i \in \mathbb{N}} \\ \text{converges whenever } f \in C_0(X) \text{ and } x \in X \end{array} \right\}.$$

Note that, in terms of $\mathcal{A}$, Proposition 4.2.2 states that if $(S,T)$ is a $C_0(X)$-equicontinuous Markov–Feller pair, then for every strictly increasing sequence $(n_k)_{k \in \mathbb{N}}$ of natural numbers there exists a subsequence $(n_{k_l})_{l \in \mathbb{N}}$ of $(n_k)_{k \in \mathbb{N}}$ such that $(n_{k_l})_{l \in \mathbb{N}}$, as a sequence of natural numbers, belongs to $\mathcal{A}$.

For every $\alpha \in \mathcal{A}$, $\alpha = (n_i)_{i \in \mathbb{N}}$ and $f \in C_0(X)$, let $f_\alpha : X \to \mathbb{R}$ be defined by $f_\alpha(x) = \lim_{i \to \infty} A_{n_i}(S)f(x)$ for every $x \in X$ (obviously, $f_\alpha$ is well defined (in the sense that the limit defining $f_\alpha(x)$ exists for every $x \in X$) since $\alpha \in \mathcal{A}$).

**Proposition 4.2.3.** *Assume that the Markov–Feller pair $(S,T)$ is $C_0(X)$-equiconti-nuous, let $\alpha \in \mathcal{A}$, and let $f \in C_0(X)$. Then $f_\alpha \in C_b(X)$ and $Sf_\alpha = f_\alpha$.*

*Proof.* Let $\alpha \in \mathcal{A}$, $\alpha = (n_i)_{i \in \mathbb{N}}$, and let $f \in C_0(X)$. Since $S$ is a contraction of $C_b(X)$, it is easy to see that $f_\alpha$ is a bounded function.

We now prove that $f_\alpha$ is continuous. To this end, let $(x_j)_{j \in \mathbb{N}}$ be a convergent sequence of elements of $X$, let $x = \lim_{j \to \infty} x_j$, and let $\varepsilon \in \mathbb{R}$, $\varepsilon > 0$. Since $S$ is $C_0(X)$-equicontinuous, there exists $j_\varepsilon \in \mathbb{N}$ such that $|S^m f(x_j) - S^m f(x)| < \frac{\varepsilon}{2}$ for every $j \geq j_\varepsilon$ and $m \in \mathbb{N} \cup \{0\}$. It follows that

$$|f_\alpha(x_j) - f_\alpha(x)| = \left| \lim_{i \to \infty} \frac{1}{n_i} \sum_{m=0}^{n_i-1} (S^m f(x_j) - S^m f(x)) \right|$$

$$= \lim_{i \to \infty} \frac{1}{n_i} \sum_{m=0}^{n_i-1} |S^m f(x_j) - S^m f(x)| \leq \frac{\varepsilon}{2}$$

for every $j \geq j_\varepsilon$. Thus, $(f_\alpha(x_j))_{j \in \mathbb{N}}$ converges to $f_\alpha(x)$. Therefore, $f_\alpha$ is a contin-uous function.

Finally, we prove that $Sf_\alpha = f_\alpha$. Using the Lebesgue dominated convergence theorem, we obtain that

$$|Sf_\alpha(x) - f_\alpha(x)| = |\langle Sf_\alpha - f_\alpha, \delta_x \rangle| = |\langle f_\alpha, T\delta_x \rangle - \langle f_\alpha, \delta_x \rangle|$$

$$= \left| \left\langle \lim_{i \to \infty} A_{n_i}(S)f, T\delta_x \right\rangle - \left\langle \lim_{i \to \infty} A_{n_i}(S)f, \delta_x \right\rangle \right| = \left| \lim_{i \to \infty} \langle A_{n_i}(S)f, T\delta_x - \delta_x \rangle \right|$$

$$= \left| \lim_{i \to \infty} \left\langle f, \frac{1}{n_i} \sum_{m=0}^{n_i - 1} T^m(T\delta_x - \delta_x) \right\rangle \right| = \left| \lim_{i \to \infty} \left\langle f, \frac{T^{n_i}\delta_x - \delta_x}{n_i} \right\rangle \right| = 0$$

for every $x \in X$. $\qquad\square$

If $\alpha \in \mathcal{A}$, $\alpha = (n_i)_{i \in \mathbb{N}}$, and if $\mu \in \mathcal{M}(X)$, then we can define a map $\mu_\alpha : C_0(X) \to \mathbb{R}$ as follows: $\mu_\alpha(f) = \lim_{i \to \infty} \langle A_{n_i}(S)f, \mu \rangle$ for every $f \in C_0(X)$. Clearly, $\mu_\alpha$ is well defined (in the sense that the sequence $(\langle A_{n_i}(S)f, \mu \rangle)_{i \in \mathbb{N}}$ is convergent whenever $f \in C_0(X)$) since $\alpha \in \mathcal{A}$, and it is easy to see that $\mu_\alpha$ is linear. Clearly, $\mu_\alpha$ is positive whenever $\mu \geq 0$; hence, $\mu_\alpha \in \mathcal{M}(X)$ (because $\mu_\alpha$ is continuous) whenever $\mu \geq 0$. In general, $\mu_\alpha \in \mathcal{M}(X)$ even if $\mu$ is not necessarily positive because $\mu = \mu^+ - \mu^-$; so, $\mu_\alpha = \mu_\alpha^+ - \mu_\alpha^-$, and $\mu_\alpha^+$ and $\mu_\alpha^-$ are elements of $\mathcal{M}(X)$.

**Proposition 4.2.4.** *Let $\alpha \in \mathcal{A}$, $\alpha = (n_i)_{i \in \mathbb{N}}$, and let $\mu \in \mathcal{M}(X)$. Then $T\mu_\alpha = \mu_\alpha$.*

*Proof.* Let $\alpha \in \mathcal{A}$, $\alpha = (n_i)_{i \in \mathbb{N}}$, and let $\mu \in \mathcal{M}(X)$. Since $\mu_\alpha = \mu_\alpha^+ - \mu_\alpha^-$, and since $(a\mu)_\alpha = a\mu_\alpha$ for every $a \in \mathbb{R}$, it is enough to prove the proposition under the assumption that $\mu \geq 0$ and $\|\mu\| = 1$ (that is, under the assumption that $\mu$ is a probability measure on $(X, \mathcal{B}(X))$). Thus, assume that $\mu$ is a probability.

In order to prove that $T\mu_\alpha = \mu_\alpha$, we will use the Lasota–Yorke lemma (Theorem 1.2.4). To this end, let $L$ be a Banach limit, and define $\phi : C_b(X) \to \mathbb{R}$, $\phi(f) = L\left(((\langle f, A_{n_i}(T)\mu \rangle)_{i \in \mathbb{N}}\right)$ for every $f \in C_b(X)$. (Note that, since for every $f \in C_b(X)$, the sequence $(\langle f, A_{n_i}(T)\mu \rangle)_{i \in \mathbb{N}}$ is bounded, the Banach limit can be applied to the sequence, so $\phi$ is well-defined.) It is easy to see that the restriction of $\phi$ to $C_0(X)$ is $\mu_\alpha$. In view of Theorem 1.2.4, in order to complete the proof of the proposition, we have to show that $\phi(Sf) = \phi(f)$ for every $f \in C_0(X)$. (We also have to prove that $\phi(1_X) = 1$, but this is obvious.)

Thus, let $f \in C_0(X)$. Then

$$|\phi(Sf) - \phi(f)| = \left| L\left(((\langle Sf, A_{n_i}(T)\mu \rangle)_{i \in \mathbb{N}}\right) - L\left(((\langle f, A_{n_i}(T)\mu \rangle)_{i \in \mathbb{N}}\right) \right|$$

$$= \left| L\left(((\langle f, TA_{n_i}(T)\mu \rangle)_{i \in \mathbb{N}}\right) - L\left(((\langle f, A_{n_i}(T)\mu \rangle)_{i \in \mathbb{N}}\right) \right|$$

$$= \left| L\left(((\langle f, TA_{n_i}(T)\mu - A_{n_i}(T)\mu \rangle)_{i \in \mathbb{N}}\right) \right|$$

$$= \left| L\left(\left(\left\langle f, \frac{T^{n_i}\mu - \mu}{n_i} \right\rangle\right)_{i \in \mathbb{N}}\right) \right| = 0.$$

The last equality holds true because the sequence $\left(\left\langle f, \dfrac{T^{n_i}\mu - \mu}{n_i} \right\rangle\right)_{i \in \mathbb{N}}$ converges

to zero, so $L\left(\left(\left\langle f, \dfrac{T^{n_i}\mu - \mu}{n_i} \right\rangle\right)_{i \in \mathbb{N}}\right) = 0$. $\qquad\square$

## 4.3   Mean Ergodic Theorems

As mentioned earlier, our goal in this section is to prove a weak* mean ergodic theorem for $C_0(X)$-equicontinuous Markov–Feller pairs. Also in this section, we will discuss several consequences of the theorem, and some examples.

Like in the previous section, we assume given a Markov–Feller pair $(S, T)$ defined on a locally compact separable metric space $(X, d)$.

**Theorem 4.3.1 (Weak* Mean Ergodic Theorem).** *Assume that $(S, T)$ is a $C_0(X)$-equicontinuous Markov–Feller pair. Then, for every $\mu \in \mathcal{M}(X)$, the sequence*

$$\left( \frac{1}{n} \sum_{m=0}^{n-1} T^m \mu \right)_{n \in \mathbb{N}} \quad converges \ in \ the \ weak^* \ topology \ of \ \mathcal{M}(X); \ if \ \mu^* \ is \ the \ weak^*$$

*limit of* $\left( \dfrac{1}{n} \displaystyle\sum_{m=0}^{n-1} T^m \mu \right)_{n \in \mathbb{N}}$ *, then* $T\mu^* = \mu^*$*; also,* $\mu^* \geq 0$ *whenever* $\mu \geq 0$.

*Proof.* Assume that $(S, T)$ is a $C_0(X)$-equicontinuous Markov–Feller pair. Clearly, in order to prove that $\left( \dfrac{1}{n} \displaystyle\sum_{m=0}^{n-1} T^m \mu \right)_{n \in \mathbb{N}}$ converges in the weak* topology of $\mathcal{M}(X)$ whenever $\mu \in \mathcal{M}(X)$, we may and do assume that $\mu \geq 0$ and $\|\mu\| = 1$ (we can make the assumption that $\mu$ is a probability, because, if $\mu = 0$ the convergence is obvious, and because $\left( \dfrac{1}{n} \displaystyle\sum_{m=0}^{n-1} T^m (a_1 \nu_1 + a_2 \nu_2) \right)_{n \in \mathbb{N}}$ weak* converges whenever $a_i \in \mathbb{R}$, $i = 1, 2$, and $\nu_i \in \mathcal{M}(X)$, $i = 1, 2$ are such that $\left( \dfrac{1}{n} \displaystyle\sum_{m=0}^{n-1} T^m \nu_i \right)_{n \in \mathbb{N}}$ weak* converges for every $i = 1, 2$).

Let $\mathcal{A}$ be the subset of $\mathbb{N}^{\mathbb{N}}$ defined in Section 4.2 ($\mathcal{A}$ is the set of all strictly increasing sequences $(n_i)_{i \in \mathbb{N}}$ of natural numbers such that the sequence $(A_{n_i}(S)f(x))_{i \in \mathbb{N}}$ converges whenever $f \in C_0(X)$ and $x \in X$). Using the comments made before Proposition 4.2.4, we obtain that for every $\alpha \in \mathcal{A}$, $\alpha = (n_i)_{i \in \mathbb{N}}$, the sequence $(A_{n_i}(T)\mu)_{i \in \mathbb{N}}$ converges in the weak* topology of $\mathcal{M}(X)$ to $\mu_\alpha$ (where $\mu_\alpha$ is the element of $\mathcal{M}(X)$ defined in the above-mentioned comments) since $\langle A_{n_i}(S)f, \mu \rangle = \langle f, A_{n_i}(T)\mu \rangle$ for every $f \in C_0(X)$ and $i \in \mathbb{N}$. Taking into consideration that (by Proposition 4.2.2) for every strictly increasing sequence $(n_k)_{k \in \mathbb{N}}$ of natural numbers, there exists a subsequence $(n_{k_l})_{l \in \mathbb{N}}$ of $(n_k)_{k \in \mathbb{N}}$ such that $(n_{k_l})_{l \in \mathbb{N}}$, as a sequence of natural numbers, belongs to $\mathcal{A}$, and using Proposition 4.2.1, we obtain that, in order to prove the weak* convergence of $(A_n(T)\mu)_{n \in \mathbb{N}}$, it is enough to prove that $\mu_\alpha = \mu_\beta$ whenever $\alpha \in \mathcal{A}$ and $\beta \in \mathcal{A}$.

Thus, let $\alpha \in \mathcal{A}$, $\alpha = (n_i)_{i \in \mathbb{N}}$ and $\beta \in \mathcal{A}$, $\beta = (n_i')_{i \in \mathbb{N}}$. We have to prove that

$$\mu_\alpha \leq \mu_\beta \tag{4.3.1}$$

and that

$$\mu_\beta \leq \mu_\alpha. \tag{4.3.2}$$

However, it is enough to prove only (4.3.1) because the proof of (4.3.2) is obtained by switching the roles of $\alpha$ and $\beta$ in the proof of (4.3.1).

In order to prove that $\mu_\alpha \leq \mu_\beta$, let $\tilde{\mu}_\alpha : C_b(X) \to \mathbb{R}$ be defined by $\tilde{\mu}_\alpha(g) = \int g(x)\,\mathrm{d}\mu_\alpha(x)$ for every $g \in C_b(X)$. Note that $\tilde{\mu}_\alpha$ is the standard extension of $\mu_\alpha$ (see the subsection *The Lasota–Yorke Lemma* of Section 1.2). Also, let $L$ be a Banach limit, and let $\mu_\alpha^* : C_b(X) \to \mathbb{R}$ be defined by $\mu_\alpha^*(g) = L\left(\left(\langle g, A_{n_i}(T)\mu\rangle\right)_{i\in\mathbb{N}}\right)$ for every $g \in C_b(X)$. Clearly, $\mu_\alpha^*$ is a positive (bounded) linear functional. It is also easy to see that the restriction of $\mu_\alpha^*$ to $C_0(X)$ is $\mu_\alpha$; thus, using an argument offered in the proof of Theorem 1.2.4 (Lasota–Yorke lemma), we obtain that

$$\tilde{\mu}_\alpha(g) \leq \mu_\alpha^*(g) \tag{4.3.3}$$

for every $g \in C_b(X)$, $g \geq 0$.

Clearly, the proof of the inequality (4.3.1) will be completed if we prove that $\langle f, \mu_\alpha \rangle \leq \langle f, \mu_\beta \rangle$ whenever $f \in C_0(X)$, $f \geq 0$.

Thus, let $f \in C_0(X)$, $f \geq 0$. Since (by Proposition 4.2.4) $T\mu_\alpha = \mu_\alpha$, it follows that $\langle f, \mu_\alpha \rangle = \left\langle f, A_{n_i'}(T)\mu_\alpha \right\rangle = \left\langle A_{n_i'}(S)f, \mu_\alpha \right\rangle$ for every $i \in \mathbb{N}$. Since $\beta \in \mathcal{A}$, the sequence $\left( A_{n_i'}(S)f \right)_{i\in\mathbb{N}}$ converges pointwise. Let $f_\beta : X \to \mathbb{R}$ be the pointwise limit of $\left( A_{n_i'}(S)f \right)_{i\in\mathbb{N}}$. By Proposition 4.2.3 the function $f_\beta$ belongs to $C_b(X)$. Using the Lebesgue dominated convergence theorem, we obtain that $\tilde{\mu}_\alpha(f_\beta) = \lim\limits_{i\to\infty} \left\langle A_{n_i'}(S)f, \mu_\alpha \right\rangle = \langle f, \mu_\alpha \rangle$; consequently, by (4.3.3), we get

$$\langle f, \mu_\alpha \rangle \leq \mu_\alpha^*(f_\beta) = L\left(\left(\langle f_\beta, A_{n_i}(T)\mu\rangle\right)_{i\in\mathbb{N}}\right). \tag{4.3.4}$$

Proposition 4.2.3 implies that $Sf_\beta = f_\beta$; hence

$$\langle f_\beta, A_{n_i}(T)\mu \rangle = \langle A_{n_i}(S)f_\beta, \mu \rangle = \langle f_\beta, \mu \rangle$$

for every $i \in \mathbb{N}$. Thus, all the terms of the sequence $(\langle f_\beta, A_{n_i}(T)\mu\rangle)_{i\in\mathbb{N}}$ are equal to $\langle f_\beta, \mu \rangle$; therefore, it follows that

$$L\left(\left(\langle f_\beta, A_{n_i}(T)\mu\rangle\right)_{i\in\mathbb{N}}\right) = \langle f_\beta, \mu \rangle. \tag{4.3.5}$$

Since $f_\beta$ is the pointwise limit of the sequence $\left( A_{n_i'}(S)f \right)_{i\in\mathbb{N}}$, using the Lebesgue dominated convergence theorem, and taking into consideration that $\mu_\beta$ is the weak* limit of the sequence $\left( A_{n_i'}(T)\mu \right)_{i\in\mathbb{N}}$, we obtain that

$$\langle f_\beta, \mu \rangle = \lim_{i\to\infty} \left\langle A_{n_i'}(S)f, \mu \right\rangle = \lim_{i\to\infty} \left\langle f, A_{n_i'}(T)\mu \right\rangle = \langle f, \mu_\beta \rangle. \tag{4.3.6}$$

Clearly, (4.3.4), (4.3.5), and (4.3.6) imply that $\langle f, \mu_\alpha \rangle \leq \langle f, \mu_\beta \rangle$.

We have therefore proved that $\left( \dfrac{1}{n} \sum\limits_{m=0}^{n-1} T^m \mu \right)_{n\in\mathbb{N}}$ converges in the weak* topology of $\mathcal{M}(X)$ whenever $\mu \in \mathcal{M}(X)$.

Now let $\mu \in \mathcal{M}(X)$, and let $\mu^* \in \mathcal{M}(X)$ be the weak* limit of the sequence $(A_n(T)\mu)_{n\in\mathbb{N}}$. Then Proposition 4.2.2 tells us that the collection $\mathcal{A}$ of subsequences is nonempty (actually, from our discussion so far, we can infer that $\mathcal{A}$ is the collection of all strictly increasing sequences of natural numbers (see Corollary 4.3.2 below)). In order to prove that $(A_n(T)\mu)_{n\in\mathbb{N}}$ weak* converges to $\mu^*$ we have proved that for every $\alpha \in \mathcal{A}$, $\alpha = (n_i)_{i\in\mathbb{N}}$, the sequence $(A_{n_i}(T)\mu)_{i\in\mathbb{N}}$ weak* converges to $\mu^*$. By Proposition 4.2.4 we obtain that $T\mu^* = \mu^*$.

Finally, if $\mu \geq 0$, and $\mu^*$ is the weak* limit of $(A_n(T)\mu)_{n\in\mathbb{N}}$, then it is easy to see that $\mu^* \geq 0$ (since $\langle f, A_n(T)\mu \rangle \geq 0$ whenever $f \in C_0(X)$, $f \geq 0$, and $n \in \mathbb{N}$). $\qquad\square$

Theorem 4.3.1 has the following consequence:

**Corollary 4.3.2 (Pointwise Mean Ergodic Theorem).** *Assume that the Markov–Feller pair $(S,T)$ is $C_0(X)$-equicontinuous. If $f \in C_0(X)$, then $(A_n(S)f)_{n\in\mathbb{N}}$ converges pointwise; that is, $(A_n(S)f(x))_{n\in\mathbb{N}}$ converges whenever $x \in X$.*

*Proof.* Let $f \in C_0(X)$ and $x \in X$. Since

$$A_n(S)f(x) = \langle A_n(S)f, \delta_x \rangle = \langle f, A_n(T)\delta_x \rangle$$

for every $n \in \mathbb{N}$, and since (by Theorem 4.3.1) the sequence $(A_n(T)\delta_x)_{n\in\mathbb{N}}$ weak* converges, it follows that $(A_n(S)f(x))_{n\in\mathbb{N}}$ converges. $\qquad\square$

The above corollary can be restated as follows:

**Corollary 4.3.3.** *Assume that the Markov–Feller pair $(S,T)$ is $C_0(X)$-equicontinuous. Then the sequence $(\langle A_n(S)f, \mu \rangle)_{n\in\mathbb{N}\cup\{0\}}$ converges whenever $f \in C_0(X)$ and $\mu \in \mathcal{M}(X)$.*

*Proof.* Let $f \in C_0(X)$ and $\mu \in \mathcal{M}(X)$. Clearly, we may assume that $\mu \geq 0$. Since the sequence $(A_n(S)f)_{n\in\mathbb{N}\cup\{0\}}$ is bounded and (by Corollary 4.3.2) converges pointwise, using the Lebesgue dominated convergence theorem, we obtain that $(\langle A_n(S)f, \mu \rangle)_{n\in\mathbb{N}\cup\{0\}}$ converges. $\qquad\square$

Theorem 4.3.1, Corollary 4.3.2, and Corollary 4.3.3 can be thought of as extensions of Theorem 1 and Theorem 5.3.1 of M. Rosenblatt [59] and [60], respectively. In the above-mentioned two theorems, Rosenblatt has considered the case of a compact space $X$, and his results imply that if $(S,T)$ is a Markov–Feller pair defined on $X$, then the sequence $(A_n(S)f)_{n\in\mathbb{N}}$ converges in the norm topology of $C_b(X)$ $(= C_0(X))$ whenever $f \in C_b(X)$ $(= C_0(X))$. We will see below that in the locally compact case we cannot expect norm convergence of $(A_n(S)f)_{n\in\mathbb{N}}$; also, we cannot expect that $(A_n(S)f)_{n\in\mathbb{N}}$ converges pointwise whenever $f \in C_b(X)$.

The next theorem summarizes Theorem 4.3.1 and Corollary 4.3.2 in the case in which we deal with a Markov–Feller pair defined on a discrete space.

**Theorem 4.3.4.** *Assume that $(S,T)$ is a Markov–Feller pair defined on $(X,d)$, and that $(X,d)$ is discrete. Then:*

(a) *The sequence $(A_n(T)\mu)_{n\in\mathbb{N}}$ converges in the weak\* topology of $\mathcal{M}(X)$ whenever $\mu \in \mathcal{M}(X)$.*

(b) *The sequence $(A_n(T)\mu(\{x\}))_{n\in\mathbb{N}}$ converges for every $\mu \in \mathcal{M}(X)$ and $x \in X$.*

(c) *The sequence $(A_n(S)f)_{n\in\mathbb{N}}$ converges pointwise for every $f \in C_0(X)$.*

*Proof.* As pointed out in Section 4.1 (after Example 4.1.3), any Markov–Feller pair defined on a discrete space is $C_0(X)$-equicontinuous. Thus, Theorem 4.3.1 and Corollary 4.3.2 imply that (a) and (c) are true.

Now let $x \in X$. Since $X$ is discrete, it follows that $1_{\{x\}}$ is an element of $C_0(X)$. Since $\langle 1_{\{x\}}, A_n(T)\mu \rangle = A_n(T)\mu(\{x\})$ for every $n \in \mathbb{N}$, using (a) we obtain that $(A_n(T)\mu(\{x\}))_{n\in\mathbb{N}}$ converges. $\qquad\square$

If $X = \mathbb{N}$ and $d$ is the usual metric on $\mathbb{N}$, then $\mathcal{M}(\mathbb{N})$, $C_b(\mathbb{N})$, and $C_0(\mathbb{N})$ can be identified with $l^1$, $l^\infty$, and $c_0$, respectively. As pointed out in Example 1.1.9, any Markov operator $T : l^1 \to l^1$ is a Markov–Feller operator, and if $S : l^\infty \to l^\infty$ is the dual of $T$, then $(S,T)$ is a Markov–Feller pair. The next theorem is a consequence of Theorem 4.3.4 in the case in which $X = \mathbb{N}$.

**Theorem 4.3.5.** *Assume that $X = \mathbb{N}$ and $\mathbb{N}$ is endowed with the usual metric, and let $(S,T)$ be a Markov–Feller pair defined on $\mathbb{N}$. If $(a_k)_{k\in\mathbb{N}} \in l^1$ and $(b_k)_{k\in\mathbb{N}} \in c_0$, then the sequences $(A_n(T)((a_k)_{k\in\mathbb{N}}))_{n\in\mathbb{N}}$ and $(A_n(S)((b_k)_{k\in\mathbb{N}}))_{n\in\mathbb{N}}$ converge pointwise (coordinatewise) in the sense that for every $j \in \mathbb{N}$ the sequences of the $j$-th coordinates of $(A_n(T)((a_k)_{k\in\mathbb{N}}))_{n\in\mathbb{N}}$ and of $(A_n(S)((b_k)_{k\in\mathbb{N}}))_{n\in\mathbb{N}}$ converge; the pointwise limit of $(A_n(T)((a_k)_{k\in\mathbb{N}}))_{n\in\mathbb{N}}$ is an element of $l^1$, and the pointwise limit of $(A_n(S)((b_k)_{k\in\mathbb{N}}))_{n\in\mathbb{N}}$ is an element of $l^\infty$.*

*Proof.* For every $j \in \mathbb{N}$ the function $1_{\{j\}}$ belongs to $c_0$, and $\langle 1_{\{j\}}, A_n(T)((a_k)_{k\in\mathbb{N}}) \rangle$ is the $j$th coordinate of $A_n(T)((a_k)_{k\in\mathbb{N}})$ for every $n \in \mathbb{N}$; since $(a)$ of Theorem 4.3.4 implies that $A_n(T)((a_k)_{k\in\mathbb{N}})$ is weak\* convergent, it follows that the sequence formed of the $j$th coordinates of $(A_n(T)((a_k)_{k\in\mathbb{N}}))_{n\in\mathbb{N}}$ is convergent. Thus, $(A_n(T)((a_k)_{k\in\mathbb{N}}))_{n\in\mathbb{N}}$ converges pointwise, and, since the weak\* limit is equal to the pointwise limit, we obtain that the pointwise limit of $(A_n(T)((a_k)_{k\in\mathbb{N}}))_{n\in\mathbb{N}}$ belongs to $l^1$.

Clearly, (c) of Theorem 4.3.4 implies that $(A_n(S)((b_k)_{k\in\mathbb{N}}))_{n\in\mathbb{N}}$ converges pointwise. Since $S$ is a positive contraction of $l^\infty$, it follows that the pointwise limit of $(A_n(S)((b_k)_{k\in\mathbb{N}}))_{n\in\mathbb{N}}$ is an element of $l^\infty$. $\qquad\square$

We mentioned earlier that, in general, if $(S,T)$ is a $C_0(X)$-equicontinuous Markov–Feller pair defined on a locally compact separable metric space $(X,d)$, then we cannot expect that $(A_n(S)f)_{n\in\mathbb{N}}$ converges in the norm of $C_b(X)$ whenever $f \in C_0(X)$. Even if $X = \mathbb{N}$, still, one can find a Markov–Feller pair $(S,T)$ and $f \in C_0(X)$ $(= c_0)$ such that $(A_n(S)f)_{n\in\mathbb{N}}$ does not norm converge. To be specific, let $(S,T)$ be the Markov–Feller pair of Example 2.1.4, and set $e_1 = (1,0,0,\ldots,0,\ldots)$. Clearly, $e_1 \in c_0$, and it is easy to see that $(A_n(S)e_1)_{n\in\mathbb{N}}$ does not converge in the norm topology of $l^\infty$, even though the sequence converges pointwise (to $(1,1,1,\ldots,1,\ldots)$). Note also that the pointwise limit of $(A_n(S)e_1)_{n\in\mathbb{N}}$

does not belong to $c_0$ even though $S(c_0) \subseteq c_0$; thus, we see that the Markov–Feller pair of Example 2.1.4 can also be used to illustrate that if $(S, T)$ is a Markov–Feller pair defined on a locally compact separable metric space $(X, d)$ such that $S(C_0(X)) \subseteq C_0(X)$, and such that $S$ is $C_0(X)$-equicontinuous, it does not necessarily follow that for every $f \in C_0(X)$, the pointwise limit of $(A_n(S)f)_{n \in \mathbb{N}}$ belongs to $C_0(X)$ (note that using Proposition 4.2.3 we can easily see that the pointwise limit of $(A_n(S)f)_{n \in \mathbb{N}}$ belongs to $C_b(X)$).

Note that if $(S, T)$ is the Markov–Feller pair of Example 2.1.4, the sequence $(A_n(S)((b_k)_{k \in \mathbb{N}}))_{n \in \mathbb{N}}$ converges pointwise (to $(b_1, b_1, b_1, \dots, b_1, \dots)$) whenever $(b_k)_{k \in \mathbb{N}} \in l^\infty$. Thus, the Markov–Feller pair of Example 2.1.4 is an example of a $C_0(X)$-equicontinuous Markov–Feller pair $(S, T)$ defined on a locally compact separable metric space $(X, d)$ such that $(A_n(S)f)_{n \in \mathbb{N}}$ converges pointwise for every $f \in C_b(X)$. However (as pointed out earlier), in general, we cannot expect the pointwise convergence of $(A_n(S)f)_{n \in \mathbb{N}}$ for every $C_0(X)$-equicontinuous Markov–Feller pair $(S, T)$ defined on a locally compact separable metric space $(X, d)$ (even if $X = \mathbb{N}$) and for every $f \in C_b(X)$. Indeed, let $(S, T)$ be the Markov–Feller pair of Example 1.1.10. Also, let $(b_k)_{k \in \mathbb{N}}$ be a sequence of real numbers such that $b_k = 0$ or 1 for every $k \in \mathbb{N}$, and such that the sequence $\left( \dfrac{1}{n} \sum_{k=1}^{n+1} b_k \right)_{n \in \mathbb{N}}$ diverges (clearly, such a sequence $(b_k)_{k \in \mathbb{N}}$ exists and is easy to construct). Obviously, $(b_k)_{k \in \mathbb{N}} \in l^\infty$, and it is easy to see that the first coordinate of $\dfrac{1}{n} \sum_{l=0}^{n-1} S^l((b_k)_{k \in \mathbb{N}})$ is $\dfrac{1}{n} \sum_{k=1}^{n+1} b_k$ for every $n \in \mathbb{N}$; hence, $\left( \dfrac{1}{n} \sum_{l=0}^{n-1} S^l((b_k)_{k \in \mathbb{N}}) \right)_{n \in \mathbb{N}}$ does not converge pointwise. Note that this Markov–Feller pair illustrates also the fact that if a Markov–Feller pair $(S, T)$ satisfies the conditions of Theorem 4.3.1 (or Theorem 4.3.4, or Theorem 4.3.5), we cannot hope for the norm convergence of $\left( \dfrac{1}{n} \sum_{k=0}^{n-1} T^k \mu \right)_{n \in \mathbb{N}}$ for every $\mu \in \mathcal{M}(X)$; indeed, if $(S, T)$ is the Markov–Feller pair of Example 1.1.10, then it is easy to see that $\left( \dfrac{1}{n} \sum_{k=0}^{n-1} T^k((a_l)_{l \in \mathbb{N}}) \right)_{n \in \mathbb{N}}$ converges pointwise to zero, but does not norm converge whenever $(a_l)_{l \in \mathbb{N}} \in l^1$, $a_l \geq 0$ for every $l \in \mathbb{N}$, and $\sum_{l=1}^{\infty} a_l = 1$.

Since, in general, the sequence $(A_n(S)f)_{n \in \mathbb{N}}$ does not converge pointwise whenever $(S, T)$ is a $C_0(X)$-equicontinuous Markov–Feller pair defined on a locally compact separable metric space $(X, d)$, and $f \in C_b(X)$, it is sometimes of interest to replace the pointwise convergence of $(A_n(S)f)_{n \in \mathbb{N}}$ for every $f \in C_b(X)$ by a weaker form of convergence. For example, given a probability $\mu$, $\mu \in \mathcal{M}(X)$, one may ask what conditions should be imposed on $\mu$ in order to guarantee that the sequence $(\langle A_n(S)f, \mu \rangle)_{n \in \mathbb{N}}$ converges to $\langle f, \mu^* \rangle$ whenever $f \in C_b(X)$, where $\mu^*$

is the weak* limit of $(A_n(T)\mu)_{n\in\mathbb{N}}$ (the existence of $\mu^*$ is assured by Theorem 4.3.1). We will discuss now such a condition. As usual (see, for example, Högnäs and Mukherjea [29]) we say that a set $\mathcal{L}$ of probabilities, $\mathcal{L} \subseteq \mathcal{M}(X)$ is *tight* if for every $\varepsilon \in \mathbb{R}$, $\varepsilon > 0$ there exists a compact subset $K$ of $X$ such that $\mu(X \setminus K) < \varepsilon$ for every $\mu \in \mathcal{L}$. We say that a sequence $(\mu_n)_{n\in\mathbb{N}}$ of probabilities, $\mu_n \in \mathcal{M}(X)$ for every $n \in \mathbb{N}$ is *tight* if the range $\{\mu_n \mid n \in \mathbb{N}\}$ of $(\mu_n)_{n\in\mathbb{N}}$ is tight.

**Proposition 4.3.6.** *Let $(S,T)$ be a $C_0(X)$-equicontinuous Markov–Feller pair defined on a locally compact separable metric space $(X,d)$, let $\mu \in \mathcal{M}(X)$ be a probability, assume that the sequence $(A_n(T)\mu)_{n\in\mathbb{N}}$ is tight, and let $\mu^*$ be the weak\* limit of $(A_n(T)\mu)_{n\in\mathbb{N}}$ (the existence of $\mu^*$ is assured by Theorem 4.3.1). Then the sequence $(\langle f, A_n(T)\mu\rangle)_{n\in\mathbb{N}}$ converges to $\langle f, \mu^*\rangle$ whenever $f \in C_b(X)$.*

*Proof.* Clearly, in order to prove the theorem, it is enough to prove that $(\langle f, A_n(T)\mu\rangle)_{n\in\mathbb{N}}$ converges to $\langle f, \mu^*\rangle$ whenever $f \in C_b(X)$, $f \geq 0$, $\|f\| \leq 1$.

Thus, let $f \in C_b(X)$, $f \geq 0$, $\|f\| \leq 1$, and let $\varepsilon \in \mathbb{R}$, $\varepsilon > 0$.

In order to simplify the notation, set $\mu_n = A_n(T)\mu$ for every $n \in \mathbb{N}$.

Since $(\mu_n)_{n\in\mathbb{N}}$ is a tight sequence of probabilities, and since $\mu^*$ is a positive regular measure, there exists a compact subset $K$ of $X$ such that $\mu^*(X \setminus K) < \dfrac{\varepsilon}{4}$, and $\mu_n(X \setminus K) < \dfrac{\varepsilon}{4}$ for every $n \in \mathbb{N}$.

By Proposition 7.1.8, p. 199 of Cohn's book [8], there exists $h \in C_c(X)$ such that $1_K \leq h \leq 1_X$. Set $g = fh$. Then $g \in C_c(X)$ because $g$ is continuous and supp $g \subseteq$ supp $h$. Moreover, $0 \leq g \leq f$ (since $0 \leq h \leq 1$ and $f \geq 0$), and $g(x) = f(x)$ for every $x \in K$ (since $h(x) = 1$ for every $x \in K$).

Since $(\mu_n)_{n\in\mathbb{N}}$ weak* converges to $\mu^*$ and since $g \in C_0(X)$, it follows that there exists $n_\varepsilon \in \mathbb{N}$ such that $|\langle g, \mu_n\rangle - \langle g, \mu^*\rangle| < \dfrac{\varepsilon}{2}$ for every $n \geq n_\varepsilon$.

We obtain that

$$
\begin{aligned}
|\langle f, \mu_n\rangle - \langle f, \mu^*\rangle| &\leq |\langle f - g, \mu_n\rangle - \langle f - g, \mu^*\rangle| + |\langle g, \mu_n\rangle - \langle g, \mu^*\rangle| \\
&< \left| \int_{X\setminus K} (f-g)\,\mathrm{d}\mu_n - \int_{X\setminus K} (f-g)\,\mathrm{d}\mu^* \right| + \frac{\varepsilon}{2} \\
&\leq \int_{X\setminus K} (f-g)\,\mathrm{d}\mu_n + \int_{X\setminus K} (f-g)\,\mathrm{d}\mu^* + \frac{\varepsilon}{2} \\
&\leq \mu_n(X\setminus K) + \mu^*(X\setminus K) + \tfrac{\varepsilon}{2} < \tfrac{\varepsilon}{4} + \tfrac{\varepsilon}{4} + \tfrac{\varepsilon}{2} = \varepsilon
\end{aligned}
$$

for every $n \geq n_\varepsilon$.

Therefore, the sequence $(\langle f, A_n(T)\mu\rangle)_{n\in\mathbb{N}}$ converges to $\langle f, \mu^*\rangle$. $\square$

Recall (see the comments made after the observation following Theorem 3.2.4 in Section 3.2) that a Markov–Feller pair $(S,T)$ defined on a locally compact separable metric space $(X,d)$ is said to be *weak\* mean ergodic* if the sequence of averages $(A_n(T)\mu)_{n\in\mathbb{N}}$ weak* converges whenever $\mu \in \mathcal{M}(X)$. Thus, in terms of weak* mean ergodicity Theorem 4.3.1 states that if $(S,T)$ is $C_0(X)$-equicontinuous, then

$(S, T)$ is weak* mean ergodic (and for every $\mu \in \mathcal{M}(X)$, the weak* limit of $(A_n(T)\mu)_{n \in \mathbb{N}}$ is $T$-invariant). Therefore, the class of weak* mean ergodic Markov–Feller pairs is larger than the class of $C_0(X)$-equicontinuous Markov–Feller pairs, so (as pointed out in the above-mentioned comments in Section 3.2) the class of weak* mean ergodic Markov–Feller pairs is fairly large.

Assume that $(S, T)$ is a $C_0(X)$-equicontinuous Markov–Feller pair, and, for every $f \in C_0(X)$, let $f^* : X \to \mathbb{R}$ be the pointwise limit of $(A_n(S)f)_{n \in \mathbb{N}}$ (the limit exists by Corollary 4.3.2; moreover, using Proposition 4.2.2, the remarks preceding Proposition 4.2.3, and Proposition 4.2.3, we obtain that $f^* \in C_b(X)$). If

$$\mathcal{A} = \left\{ f^* \in C_b(X) \ \middle| \ \begin{array}{l} f^* \text{ is the pointwise limit of } (A_n(S)f)_{n \in \mathbb{N}} \\ \text{for some } f \in C_0(X) \end{array} \right\},$$

then, by Corollary 3.2.5, the pair $(S, T)$ is uniquely ergodic if and only if there exists $x_0 \in X$ such that $f^*(x) \le f^*(x_0)$ for every $f^* \in \mathcal{A}$ and $x \in X$ (that is, if $\mathcal{A}$ has a common maximum at $x_0$).

Note that if $(S, T)$ is $C_0(X)$-equicontinuous and uniquely ergodic, if $\mu^*$ is the unique $T$-invariant probability, and if $X$ is not compact, then it is not true, in general, that $(A_n(T)\mu)_{n \in \mathbb{N}}$ weak* converges to $\mu^*$ whenever $\mu$ is a probability, $\mu \in \mathcal{M}(X)$. Indeed, if $(S, T)$ is the Markov–Feller pair of Example 1.1.14, then it is easy to see that $(S, T)$ is uniquely ergodic, and the unique invariant probability of $T$ is $\delta_{\{1\}}$ (which corresponds to the element $(1, 0, 0, 0, \ldots, 0, \ldots)$ of $l^1$); however, the sequence $(A_n(T)\delta_{\{2\}})_{n \in \mathbb{N}}$ weak* converges to 0 rather than $\delta_{\{1\}}$. Motivated by the above discussion, it makes sense to define a new type of Markov–Feller pairs as follows: we say that the Markov–Feller pair $(S, T)$ is *weak* uniquely mean ergodic* if there exists a probability $\mu^*$, $\mu^* \in \mathcal{M}(X)$ such that $(A_n(T)\mu)_{n \in \mathbb{N}}$ weak* converges to $\mu^*$ whenever $\mu$ is a probability, $\mu \in \mathcal{M}(X)$. Clearly, $\mu^*$ is $T$-invariant.

**Observation.** If $(\mu_n)_{n \in \mathbb{N}}$ is a sequence of probabilities ($\mu_n \in \mathcal{M}(X)$ for every $n \in \mathbb{N}$) and if $(\mu_n)_{n \in \mathbb{N}}$ weak* converges to some probability $\mu'$, $\mu' \in \mathcal{M}(X)$, then the sequence $(\mu_n)_{n \in \mathbb{N}}$ is tight. Indeed, let $\varepsilon \in \mathbb{R}$, $\varepsilon > 0$; since $\mu'$ is a regular probability measure, there exists a compact subset $K'$ of $X$ such that $\mu'(K') > 1 - \dfrac{\varepsilon}{2}$. By Proposition 7.1.8, p. 199 of Cohn [8], there exists $f \in C_c(X)$ such that $1_{K'} \le f \le 1_X$. Set $K'' = \operatorname{supp} f$. Since $(\mu_n)_{n \in \mathbb{N}}$ is weak* convergent to $\mu'$, it follows that there exists $n_\varepsilon \in \mathbb{N}$ such that $|\langle f, \mu_n \rangle - \langle f, \mu' \rangle| < \dfrac{\varepsilon}{2}$ for every $n \ge n_\varepsilon$; therefore,

$$\mu_n(K'') \ge \langle f, \mu_n \rangle > -\frac{\varepsilon}{2} + \langle f, \mu' \rangle \ge -\frac{\varepsilon}{2} + \mu'(K') > 1 - \varepsilon$$

for every $n \ge n_\varepsilon$. Since the measures $\mu_1, \mu_2, \ldots, \mu_{n_\varepsilon - 1}$ are regular, there exists a compact subset $K$ of $X$ such that $K \supseteq K''$ and $\mu_i(K) > 1 - \varepsilon$ for every $i = 1, 2, \ldots, n_\varepsilon - 1$. Clearly, in this case $\mu_n(K) > 1 - \varepsilon$ for every $n \in \mathbb{N}$. Since for every $\varepsilon \in \mathbb{R}$, $\varepsilon > 0$ there exists a compact subset $K$ of $X$ such that $\mu_n(K) > 1 - \varepsilon$ for every $n \in \mathbb{N}$, it follows that $(\mu_n)_{n \in \mathbb{N}}$ is tight.

The above remark implies that if $(S,T)$ is a weak* uniquely mean ergodic Markov–Feller pair, then the sequence $(A_n(T)\mu)_{n\in\mathbb{N}}$ is tight whenever $\mu$ is a probability, $\mu \in \mathcal{M}(X)$.

Using Proposition 4.3.6, we obtain that a Markov–Feller pair $(S,T)$ is weak* uniquely mean ergodic if and only if there exists a probability $\mu^*$, $\mu^* \in \mathcal{M}(X)$ such that the sequence $(\langle f, A_n(T)\mu\rangle)_{n\in\mathbb{N}}$ converges to $\langle f, \mu^*\rangle$ whenever $\mu$ is a probability, $\mu \in \mathcal{M}(X)$ and $f \in C_b(X)$. ∎

It is easy to see that a weak* uniquely mean ergodic Markov–Feller pair is uniquely ergodic. However, in general, the converse is not true; for example, the Markov–Feller pair of Example 1.1.14 is uniquely ergodic, but it is not weak* uniquely mean ergodic. The next theorem offers conditions under which unique ergodicity and weak* unique mean ergodicity are equivalent.

**Theorem 4.3.7.** *Let $(S,T)$ be a Markov–Feller pair defined on a locally compact separable metric space $(X,d)$, and assume that $X = \Gamma_{cp}$. Then $(S,T)$ is uniquely ergodic if and only if $(S,T)$ is weak* uniquely mean ergodic.*

*Proof.* Clearly, in view of our discussion so far, we only have to prove that if $(S,T)$ is uniquely ergodic, then $(S,T)$ is weak* uniquely mean ergodic. Thus, assume that $(S,T)$ is uniquely ergodic.

Since $X = \Gamma_{cp}$, it makes sense to consider the elements $\varepsilon_x$, $x \in X$ of $\mathcal{M}(X)$ that have been defined in the subsection *The KBBY Decomposition* of Section 1.2. In view of the definition of $\Gamma_{cp}$, it follows that $\varepsilon_x$ is a probability for every $x \in X$. By the remarks made after Theorem 2.1.1, the probabilities $\varepsilon_x$, $x \in X$ are $T$-invariant. Since $(S,T)$ is uniquely ergodic, it follows that $\varepsilon_x = \mu^*$ for every $x \in X$, where $\mu^*$ is the unique $T$-invariant probability. In view of the definition of $\varepsilon_x$, $x \in X$, we obtain that $\lim_{n\to\infty} A_n(S)f(x) = \langle f, \mu^*\rangle$ for every $x \in X$ and $f \in C_0(X)$.

Now let $\mu \in \mathcal{M}(X)$ be a probability, and let $f \in C_0(X)$. By the Lebesgue dominated convergence theorem, the sequence $(\langle A_n(S)f, \mu\rangle)_{n\in\mathbb{N}}$ converges (so, $(\langle f, A_n(T)\mu\rangle)_{n\in\mathbb{N}}$ converges, as well) and

$$\lim_{n\to\infty} \langle f, A_n(T)\mu\rangle = \lim_{n\to\infty} \int A_n(S)f(x)\,d\mu(x) = \int \langle f, \mu^*\rangle 1_X(x)\,d\mu(x) = \langle f, \mu^*\rangle$$

(the Lebesgue dominated convergence theorem can be applied since $(A_n(S)f)_{n\in\mathbb{N}}$ is a bounded sequence $((A_n(S)f)_{n\in\mathbb{N}}$ is bounded because $S$ is a contraction of $C_b(X)$) and converges pointwise (to $\langle f, \mu^*\rangle$)).

Since for every $\mu \in \mathcal{M}(X)$ and $f \in C_0(X)$ the sequence $(\langle f, A_n(T)\mu\rangle)_{n\in\mathbb{N}}$ converges, it follows that $(S,T)$ is weak* uniquely mean ergodic. □

In the case in which we deal with Markov–Feller pairs defined on compact metric spaces, Theorem 4.3.7 becomes:

**Corollary 4.3.8.** *Assume that the Markov–Feller pair $(S,T)$ is defined on a compact metric space $(X,d)$. Then $(S,T)$ is uniquely ergodic if and only if $(S,T)$ is weak* uniquely mean ergodic.*

*Proof.* Since the weak* unique mean ergodicity implies the unique ergodicity, we only have to prove that if $(S,T)$ is uniquely ergodic, then $(S,T)$ is weak* uniquely mean ergodic. To this end, assume that $(S,T)$ is uniquely ergodic, and let $\mu^*$ be the unique $T$-invariant probability. Then using Proposition 1.2 and Proposition 1.3, both on p. 178 of Krengel's book [32], we obtain that $(A_n(S)f)_{n\in\mathbb{N}}$ converges pointwise to $\langle f,\mu^*\rangle$ whenever $f \in C_0(X)$ $(= C_b(X))$; hence, $X = \Gamma_{cp}$, and we can apply Theorem 4.3.7 in order to conclude that $(S,T)$ is weak* uniquely mean ergodic.                                                                                  $\square$

Note that the above corollary can also be proved directly without using Theorem 4.3.7. Indeed, if $(S,T)$ is uniquely ergodic (and if $(X,d)$ is compact), then by Proposition 1.2 and Proposition 1.3, p. 178 of Krengel [32], the sequence $(A_n(S)f)_{n\in\mathbb{N}}$ converges not only pointwise, but also in the norm topology of $C_b(X)$ to $\langle f,\mu^*\rangle 1_X$ whenever $f \in C_0(X)$ $(= C_b(X))$, where $\mu^*$ is the unique $T$-invariant probability; so, it is easy to see that $(A_n(T)\mu)_{n\in\mathbb{N}}$ converges in the weak* topology of $\mathcal{M}(X)$ to $\mu^*$ whenever $\mu$ is a probability, $\mu \in \mathcal{M}(X)$.

Note also that, in general, it is not true that the unique invariant probability of a weak* uniquely mean ergodic Markov–Feller pair is an attractive probability. For example, if $a \in \mathbb{R}/\mathbb{Z}$ is such that the equivalence class $a$ contains irrational numbers, then the Markov–Feller pair $(S_a, T_a)$ of Example 1.1.11 is weak* uniquely mean ergodic, but the unique invariant probability (the Haar (Lebesgue) measure on $\mathbb{R}/\mathbb{Z}$) is obviously not an attractive probability.

If $(S,T)$ is a weak* uniquely mean ergodic Markov–Feller pair defined on the not necessarily compact space $(X,d)$, and if $\mu^*$ is the unique $T$-invariant probability, then

$$\operatorname{supp} \mu^* = \bigcap_{x\in X} \overline{\mathcal{O}(x)} \tag{4.3.7}$$

where $\overline{\mathcal{O}(x)}$ is the orbit-closure of $x \in X$ (for the definition of $\overline{\mathcal{O}(x)}$, $x \in X$, see the beginning of Section 2.2).

Clearly, a straightforward application of Theorem 3.1.1 yields the equality (4.3.7) (note that (4.3.7) is slightly more general than Corollary 3.1.2). However, (4.3.7) can also be proved directly, without thinking in terms of the KBBY decomposition. Indeed, in order to prove that $\operatorname{supp} \mu^* \subseteq \bigcap_{x\in X} \overline{\mathcal{O}(x)}$, we have to prove that $z \in \overline{\mathcal{O}(x)}$ whenever $z \in \operatorname{supp} \mu^*$ and $x \in X$. So, let $z \in \operatorname{supp} \mu^*$ and $x \in X$. Since $X$ is locally compact, we can pick $\alpha \in \mathbb{R}$, $\alpha > 0$, such that $\overline{B(z,\alpha)}$ is compact. Clearly, in order to show that $z \in \overline{\mathcal{O}(x)}$, we have to prove that $\overline{\mathcal{O}(x)} \cap B\left(z,\dfrac{\alpha}{k}\right) \neq \emptyset$ for every $k \in \mathbb{N}$. So, let $k \in \mathbb{N}$ and consider the function $f_k : X \to \mathbb{R}$, $f_k(y) = d\left(y, X \setminus B\left(z,\dfrac{\alpha}{k}\right)\right)$ for every $y \in X$ (note that we used the function $f_k$ in the proof of (a) of Theorem 2.2.1). Since $f_k \in C_0(X)$ (actually, $f_k \in C_c(X)$ because $f(y) = 0$ for every $y \in X \setminus B\left(z,\dfrac{\alpha}{k}\right)$, and $\overline{B\left(z,\dfrac{\alpha}{k}\right)}$ is compact), since $(S,T)$ is weak* uniquely mean ergodic, and since $\mu^*$ is the (unique)

$T$-invariant probability, it follows that the sequence $(\langle f_k, A_n(T)\delta_x\rangle)_{n\in\mathbb{N}}$ converges to $\langle f_k, \mu^*\rangle$. Since $f_k(y) > 0$ for every $y \in B\left(z, \dfrac{\alpha}{k}\right)$, and since $z \in \operatorname{supp} \mu^*$, it follows that $\langle f_k, \mu^*\rangle > 0$. Thus, there exists $n \in \mathbb{N}$ such that $\langle f_k, A_n(T)\delta_x\rangle > 0$; hence, there exists $l \in \{0, 1, 2, \ldots, n-1\}$ such that $\langle f_k, T^l\delta_x\rangle > 0$. Since $0 \le f_k \le \dfrac{\alpha}{k}\mathbb{1}_{B\left(z, \frac{\alpha}{k}\right)}$, we obtain that $T^l\delta_x\left(B\left(z, \dfrac{\alpha}{k}\right)\right) > 0$; hence, there exists $y \in \mathcal{O}(x) \cap B\left(z, \dfrac{\alpha}{k}\right)$ since $\operatorname{supp}(T^l\delta_x) \cap B\left(z, \dfrac{\alpha}{k}\right) \ne \emptyset$. Conversely, the inclusion $\bigcap_{x\in X} \overline{\mathcal{O}(x)} \subseteq \operatorname{supp} \mu^*$ can be proved using Proposition 1.1.7 as in the proof of (b) of Theorem 2.2.1; indeed, if $x_0 \in \operatorname{supp} \mu^*$, then $\operatorname{supp} T^n\delta_{x_0} \subseteq \operatorname{supp} \mu^*$ for every $n \in \mathbb{N} \cup \{0\}$ by Proposition 1.1.7; clearly, in view of the definition of $\overline{\mathcal{O}(x_0)}$, we obtain that $\overline{\mathcal{O}(x_0)} \subseteq \operatorname{supp} \mu^*$; hence, $\bigcap_{x\in X} \overline{\mathcal{O}(x)} \subseteq \operatorname{supp} \mu^*$ (note that the direct proof of (4.3.7) allows us to infer that $\overline{\mathcal{O}(x)} = \operatorname{supp} \mu^*$ whenever $x \in \operatorname{supp} \mu^*$).

As already pointed out, the direct proof of (4.3.7) does not involve the KBBY decomposition (actually, the proof does not involve even the Lasota–Yorke lemma or the notion of ergodic measures even though the ideas of the proof have all appeared throughout our discussion so far (especially in the proof of Theorem 2.2.1)). When we obtained the direct proof of (4.3.7), we realized that in the noncompact case some Markov–Feller pairs do not have invariant probabilities (see, for instance, Example 1.1.10), while others might be uniquely ergodic, but not weak* uniquely mean ergodic (see Example 1.1.14). These observations have triggered all the results that we have discussed in this work.

# Bibliography

[1] Y. A. Abramovich and C. D. Aliprantis, *An Invitation to Operator Theory*, Grad. Studies in Math., Vol. 50, AMS, Providence, Rhode Island, 2002.

[2] C. D. Aliprantis and O. Burkinshaw, *Positive Operators*, Academic Press, Orlando, Florida, 1985.

[3] M. Barnsley, *Fractals Everywhere*, Academic Press, San Diego/London, 1988.

[4] M. F. Barnsley, S. G. Demko, J. H. Elton, and J. S. Geronimo, *Invariant measures for Markov processes arising from iterated function systems with place-dependent probabilities*, Ann. Inst. H. Poincaré, Probab. et Statist. **24**(1988), 367-394 and **25**(1989), 589-590.

[5] M. Beboutoff, *Markov chains with a compact state space*, Rec. Math. (Mat. Sbornik) N. S. **10(52)**(1942), 213-238.

[6] M. Boshernitzan, *A unique ergodicity of minimal symbolic flows with linear block growth*, J. Anal. Math., **44**(1984/85), 77-96.

[7] M. Boshernitzan, *A condition for minimal interval exchange maps to be uniquely ergodic*, Duke Math. J., **52**(1985), 723-752.

[8] D. L. Cohn, *Measure Theory*, Birkhäuser, Boston, 1980.

[9] I. P. Cornfeld, S. V. Fomin, and Ya. G. Sinai, *Ergodic Theory*, Springer, Berlin/New York, 1982.

[10] E. M. Coven, *Sequences with minimal block growth II*, Math. Systems Theory, **8**(1975), 376-382.

[11] E. M. Coven and G. A. Hedlund *Sequences with minimal block growth*, Math. Systems Theory, **7**(1973), 138-153.

[12] E. M. Coven and M. E. Paul, *Endomorphisms of irreducible subshifts of finite type*, Math. Systems Theory, **8**(1974), 167-175.

[13] S. G. Dani and S. Muralidharan, *On ergodic averages for affine lattice actions on tori*, Monatshefte für Mathematik, **96**(1983), 17-28.

[14] P. Diaconis and D. Freedman, *Iterated random functions*, SIAM Rev., 41(1999), 45-76 (electronic).

[15] N. Dunford and J. T. Schwartz, *Linear Operators, Part I: General Theory*, Wiley, New York, 1988.

[16] F. Durand, *Linearly recurrent subshifts have a finite number of non-periodic subshift factors,* Ergod. Th.&Dynam. Sys. **20**(2000), 1061-1078.

[17] K. Falconer, *Fractal Geometry: Mathematical Foundations and Applications,* John Wiley&Sons, Chichester, England, 1990.

[18] S. R. Foguel, *The Ergodic Theory of Markov Processes,* Van Nostrand Reinhold, New York/Toronto, Ontario/London, 1969.

[19] H. Furstenberg, *Strict ergodicity and transformation of the torus,* Amer. J. Math., **83**(1961), 573-601.

[20] H. Furstenberg, *Recurrence in Ergodic Theory and Combinatorial Number Theory,* Princeton Univ. Press, Princeton, New Jersey, 1981.

[21] A. Goetz, *Dynamics of piecewise isometries,* Illinois J. Math., **44**(2000), 465-478.

[22] W. H. Gottschalk and G. A. Hedlund, *Topological Dynamics,* Colloquium Publications, Vol. 36, AMS, Providence, Rhode Island, 1955.

[23] P. R. Halmos, *Measure Theory,* Springer, New York, 1974.

[24] B. R. Henry, *Escape from the unit interval under the transformation* $x \mapsto \lambda x(1 - x),$ Proc. Amer. Math. Soc. **41**(1973), 146-150.

[25] O. Hernández-Lerma and J. B. Lasserre, *Ergodic theorems and ergodic decomposition for Markov chains,* Acta Appl. Math. **54**(1998), 99-119.

[26] O. Hernández-Lerma and J. B. Lasserre, *Existence and uniqueness of fixed points for Markov operators and Markov processes,* Proc. London Math. Soc. **76**(1998), 711-736.

[27] O. Hernández-Lerma and J. B. Lasserre, *Markov Chains and Invariant Probabilities,* Birkhäuser, Basel, 2003.

[28] H. Heyer, *Probability Measures on Locally Compact Groups,* Springer, Berlin/Heidelberg/New York, 1977.

[29] G. Högnäs and A. Mukherjea, *Probability Measures on Semigroups-Convolution Products, Random Walks, and Random Matrices,* Plenum Press, New York, 1995.

[30] M. Keane, *Non-ergodic interval exchange transformations,* Israel J. Math., **26**(1977), 188-196.

[31] H. B. Keynes and D. Newton, *A "minimal," non-uniquely ergodic interval exchange transformation,* Math. Z. **148**(1976), 101-105.

[32] U. Krengel, *Ergodic Theorems,* de Gruyter, Berlin/New York, 1985.

[33] N. Krylov and N. Bogolioubov, *La théorie générale de la mesure dans son application à l'étude des systèmes de la mécanique non linéaires,* Ann. of Math., **38**(1937), 65-113.

[34] K. Kuratowski, *Topology,* Vol. 1, Academic Press and Polish Scientific Publishers, New York and Warszawa, 1966.

[35] A. Lasota and M. C. Mackey, *The extinction of slowly evolving dynamical systems,* J. Math. Biology, **10**(1980), 333-345.

[36] A. Lasota and M. C. Mackey, *Chaos, Fractals, and Noise,* Springer, New York, 1994.

[37] A. Lasota and J. Myjak, *Semifractals,* Bull. Polish Acad. Sci., Math., **44**(1996), 5-21.

[38] A. Lasota and J. Myjak, *Markov operators and fractals,* Bull. Polish Acad. Sci., Math., **45**(1997), 197-210.

[39] A. Lasota and J. Myjak, *Semifractals on Polish spaces,* Bull. Polish Acad. Sci., Math. **46**(1998), 179-196.

[40] A. Lasota and J. Myjak, *Fractals, semifractals, and Markov operators,* International J. of Bifurcation and Chaos, **9**(1999), 307-325.

[41] A. Lasota and J. Myjak, *Attractors of multifunctions,* Bull. Polish Acad. Sci., Math. **48**(2000), 319-334.

[42] A. Lasota and J. A. Yorke, *Lower bound technique for Markov operators and iterated function systems,* Random&Comput. Dynamics, **2**(1994), 41-77.

[43] T.-Y. Li and J. A. Yorke, *Period three implies chaos,* Amer. Math. Monthly, **82**(1975), 985-992.

[44] T.-Y. Li and J. A. Yorke, *The "simplest" dynamical system,* Dynamical Systems, An International Symposium, Vol. 2, pp. 203-206, Ed.: L. Cesari, J. K. Hale, and J. P. LaSalle, Academic Press, New York/San Francisco/London, 1976.

[45] M. Lin, *Conservative Markov processes on a topological space,* Israel J. Math. **8**(1970), 165-186.

[46] D. Lind and B. Marcus, *Introduction to Symbolic Dynamics and Coding,* Cambridge Univ. Press, New York, 1995.

[47] R. M. May, *Simple mathematical models with very complicated dynamics,* Nature, **261**(1976), 459-467.

[48] W. de Melo and S. van Strien, *One-Dimensional Dynamics,* Ergebnisse der Mathematik und ihrer Grenzgebiete, Band 25, Springer, Berlin/Heidelberg, 1993.

[49] P. Meyn and R. L. Tweedie, *Markov Chains and Stochastic Stability,* Springer, London, 1993.

[50] M. Misiurewicz, *Absolutely continuous measures for certain maps of an interval,* Publ. Math. IHES, **53**(1981), 17-51.

[51] E. Nummelin, *General Irreducible Markov Chains and Non-Negative Operators,* Cambridge Univ. Press, Cambridge, England, 1984.

[52] S. Orey, *Lecture Notes on Limit Theorems for Markov Chain Transition Probabilities,* Van Nostrand Reinhold, London/New York/Toronto, Ontario, 1971.

[53] J. C. Oxtoby, *Ergodic sets,* Bull. Amer. Math. Soc. **58**(1952), 116-136.

[54] J. C. Oxtoby and S. M. Ulam, *On the existence of a measure invariant under a transformation,* Ann. of Math. **40**(1939), 560-566.

[55] M. E. Paul, *Minimal symbolic flows having minimal block growth,* Math. Systems Theory, **8**(1975), 309-315.

[56] G. Pianigiani, *Existence of invariant measures for piecewise continuous transformations,* Ann. Polonici Math. **40**(1981), 39-45.

[57] D. Revuz, *Markov Chains,* North-Holland/American Elsevier, Amsterdam/New York, 1975.

[58] C. Robinson, *Dynamical Systems - Stability, Symbolic Dynamics, and Chaos,* CRC Press, Boca Raton/Ann Arbor/London/Tokyo, 1995.

[59] M. Rosenblatt, *Equicontinuous Markov operators,* Theor. Prob. Appl. **9**(1964), 205-222.

[60] M. Rosenblatt, *Markov Processes. Structure and Asymptotic Behavior,* Springer, Berlin/Heidelberg, 1971.

[61] H. L. Royden, *Real Analysis,* Third Edition, Macmillan, New York and Collier Macmillan, London, 1988.

[62] D. Ruelle, *Applications conservant une mesure absolument continue par rapport á dx sur [0, 1],* Comm. Math. Phys. **55**(1977), 47-51.

[63] H. H. Schaefer, *Banach Lattices and Positive Operators,* Springer, New York/Heidelberg/Berlin, 1974.

[64] A. V. Skorokhod, *Topologically recurrent Markov chains: ergodic properties,* Theor. Probab. Appl. **31**(1986), 563-571.

[65] L. Sucheston, *Banach limits,* Amer. Math. Monthly, **74**(1967), 308-311.

[66] T. Szarek, *Invariant Measures for Nonexpansive Markov Operators on Polish Spaces,* Dissertationes Mathematicae 425, Warsaw, 2003.

[67] P. Walters, *An Introduction to Ergodic Theory,* Springer, New York, 1982.

[68] K. Yosida, *Markov processes with a stable distribution,* Proc. Imp. Acad. Tokyo, **16**(1940), 43-48.

[69] K. Yosida, *Simple Markoff process with a locally compact phase space,* Math. Japonicae **1**(1948), 99-103.

[70] K. Yosida, *Functional Analysis,* Third Edition, Springer, New York, 1971.

[71] A. C. Zaanen, *Introduction to Operator Theory in Riesz Spaces,* Springer, Berlin/Heidelberg, 1997.

[72] R. Zaharopol, *Attractive probability measures and their supports,* to appear in Rev. Roumaine Math. Pures Appl., **49**(2004), No. 4.

[73] R. Zaharopol, *Iterated function systems generated by strict contractions and place-dependent probabilities,* Bull. Polish Acad. Sci. Math. **48**(2000), 429-438.

[74] R. Zaharopol and G. Zbaganu, *Dobrushin coefficients of ergodicity and asymptotically stable $L^1$-contractions,* J. Theoret. Probab. **12**(1999), 885-902.

# Index

# Frontiers in Mathematics

This new series is designed to be a repository for up-to-date
research results which have been prepared for a wider audience. Graduates and postgraduates as well as scientists will
benefit from the latest developments at the research frontiers in mathematics and at the "frontiers" between mathematics and other fields like computer science, physics, biology, economics, finance, etc. All volumes will be online available at SpringerLink.

## Your Specialized Publisher in Mathematics
### Birkhäuser

For orders originating from all over the world
except USA/Canada/Latin America:

Birkhäuser Verlag AG
c/o Springer GmbH & Co
Haberstrasse 7
D-69126 Heidelberg
Fax: +49 / 6221 / 345 4 229
e-mail: birkhauser@springer.de
http://www.birkhauser.ch

For orders originating in the
USA/Canada/Latin America:

Birkhäuser
333 Meadowland Parkway
USA-Secaucus
NJ 07094-2491
Fax: +1 / 201 / 348 4505
e-mail: orders@birkhauser.com

---

◼ **Thas, K.**, Ghent University, Ghent, Belgium

**Symmetry in Finite Generalized Quadrangles**

2004. 240 pages. Softcover. ISBN 3-7643-6158-1

In this book, a classification of finite generalized
quadrangles based on the possible subconfigurations of
axes of symmetry is proposed, extending thus the
celebrated Lenz-Barlotti classification for projective planes
to the theory of generalized quadrangles.
Several open problems and long-standing conjectures are
solved, respectively answered, by new techniques arising
from a mixture of geometrical, combinatorial and group
theoretical arguments. Many new, previously unpublished
results with proofs are presented.
The book is aimed at advanced graduate students and
researchers in the area. Readers will find a self-contained
introduction to the modern theory of finite generalized
quadrangles and related structures, as well as a detailed
account of the classification and its implications.

◼ **Krausshar, R.S.**, Ghent University, Ghent, Belgium

**Generalized Analytic Automorphic Forms in
Hypercomplex Spaces**

2004. 182 pages. Softcover. ISBN 3-7643-7059-9

The aim of this book is to provide a first comprehensive
overview of the basic theory of hypercomplex-analytic
automorphic forms and functions for arithmetic subgroups
of the Vahlen group in higher dimensional spaces. It gives
a summary on the research results obtained over the last
five years and establishes a new field within the theory of
functions of hypercomplex variables and within analytic
number theory.
Hypercomplex-analyticity generalizes the concept of
complex analyticity in the sense of considering
null-solutions to higher dimensional Cauchy-Riemann type
systems. Vector- and Clifford algebra-valued Eisenstein
and Poincaré series are constructed within this framework
and a detailed description of their analytic and number
theoretical properties is provided. In particular, explicit
relationships to higher dimensional vector valued variants
of the Riemann zeta function and Dirichlet series are
established.

# Frontiers in Mathematics

## Further titles

■ **Bouchut, F.**, CNRS & Ecole Normale Supérieure, Paris, France

**Nonlinear Stability of Finite Volume Methods for Hyperbolic Conservation Laws**

and Well-Balanced Schemes for Sources
2004. 144 pages. Softcover
ISBN 3-7643-6665-6

This book is devoted to finite volume methods for hyperbolic systems of conservation laws. It differs from previous expositions on the subject in that the accent is put on the development of tools and the design of schemes for which one can rigorously prove nonlinear stability properties. Sufficient conditions for a scheme to preserve an invariant domain or to satisfy discrete entropy inequalities are systematically exposed, with analysis of suitable CFL conditions.

The monograph intends to be a useful guide for the engineer or researcher who needs very practical advice on how to get such desired stability properties. The notion of approximate Riemann solver and the relaxation method, which are adapted to this aim, are especially explained. In particular, practical formulas are provided in a new variant of the HLLC solver for the gas dynamics system, taking care of contact discontinuities, entropy conditions, and including vacuum. In the second half of the book, nonconservative schemes handling source terms are analyzed in the same spirit. The recent developments on well-balanced schemes that are able to capture steady states are explained within a general framework that includes analysis of consistency and order of accuracy. Several schemes are compared for the Saint Venant problem concerning positivity and the ability to treat resonant data. In particular, the powerful and recently developed hydrostatic reconstruction method is detailed.

■ **Kasch, F.**, Universität München, Germany / **Mader, A.**, Hawaii University

**Rings, Modules, and the Total**
2004. 148 pages. Softcover
ISBN 3-7643-7125-0

In a nutshell, this monograph deals with direct decompositions of modules and associated concepts. The central notion of "partially invertible homomorphisms", namely those that are factors of a non-zero idempotent, is introduced in a very accessible fashion. Units and regular elements are partially invertible. The "total" consists of all elements that are not partially invertible. The total contains the radical and the singular and cosingular submodules, but while the total is closed under right and left multiplication, it may not be closed under addition. Cases are discussed where the total is additively closed. The total is particularly suited to deal with the endomorphism ring of a direct sum of modules that all have local endomorphism rings and is applied in this case. Further applications are given for torsion-free Abelian groups. This book offers for the first time a comprehensive and readable exposition of results on the total. Although dealing with recent research, the material is accessible to anyone with a basic knowledge of ring and module theory. A short introduction to torsion-free Abelian groups is included. The subject is by no means exhausted and topics for further research can easily be found.

■ **Zaharopol, R.**, Mathematical Reviews, Ann Arbor, USA

**Invariant Probabilities of Markov-Feller Operators and Their Supports**

2005. 128 pages. Softcover
ISBN 3-7643-7134-X

For orders originating from all over the world except USA/Canada/Latin America:

Birkhäuser Verlag AG
c/o Springer GmbH & Co
Haberstrasse 7
D-69126 Heidelberg

Fax: +49 / 6221 / 345 4 229
e-mail: birkhauser@springer.de
http://www.birkhauser.ch

For orders originating in the USA/Canada/Latin America:

Birkhäuser
333 Meadowland Parkway
USA-Secaucus
NJ 07094-2491

Fax: +1 / 201 / 348 4505
e-mail: orders@birkhauser.com
http://www.birkhauser.com

# Oberwolfach Seminars

The workshops organized by the Mathematisches Forschungsinstitut Oberwolfach are intended to introduce students and young mathematicians to current fields of research. By means of these well-organized seminars, also scientists from other fields will be introduced to new mathematical ideas.
The publication of these workshops in the series Oberwolfach Seminars (formerly DMV Seminar) makes the material available to an even larger audience.

## Your Specialized Publisher in Mathematics

## Birkhäuser

For orders originating from all over the world except USA/Canada/Latin America:
All countries excluding those listed below:
Birkhäuser Verlag AG
c/o Springer Auslieferungs-Gesellschaft (SAG)
Customer Service
Haberstrasse 7, D-69126 Heidelberg
Tel.: +49 / 6221 / 345 0
Fax: +49 / 6221 / 345 42 29
e-mail: orders@birkhauser.ch

For orders originating in the USA/Canada/Latin America:

Birkhäuser
333 Meadowland Parkway
USA-Secaucus
NJ 07094-2491
Fax: +1 201 348 4505
e-mail: orders@birkhauser.com

**OWS 33: Kreck, M. / Lück, W.**, The Novikov Conjecture: Geometry and Algebra (2004). ISBN 3-7643-7141-2

**DMV 32: Bolthausen, E. / Sznitman, A.-S.**, Ten Lectures on Random Media (2002). ISBN 3-7643-6703-2

**DMV 31: Huckleberry, A. / Wurzbacher, T.** (Eds.), Infinite Dimensional Kähler Manifolds (2001). ISBN 3-7643-6602-8

**DMV 30: Scholz, E.** (Ed.), Hermann Weyl's Raum-–Zeit—Materie and a General Introduction to His Scientific Work (2001). ISBN 3-7643-6476-9

**DMV 29: Kalai, G. / Ziegler, G.M.** (Eds.), Polytopes — Combinatorics and Computation (2000). ISBN 3-7643-6351-7

**DMV 28: Cercignani, C. / Sattinger, D.**, Scaling Limits and Models in Physical Processes (1998). ISBN 3-7643-5985-4

**DMV 27: Knauf, A. / Sinai, Y.G.**, Classical Nonintegrability, Quantum Chaos (1997). ISBN 3-7643-5708-8

**DMV 26: Miyaoka, Y. / Peternell, T.**, Geometry of Higher Dimensional Algebraic Varieties (1997). ISBN 3-7643-5490-9

**DMV 25: Ballmann, W.**, Lectures on Spaces of Nonpositive Curvature (1995). ISBN 3-7643-5242-6

**DMV 24: Aubry, M.**, Homotopy Theory and Models (1995). ISBN 3-7643-5185-3

**DMV 23: Falk, M. / Hüsler, J. / Reiss, R.-D.**, Lawas of Small Numbers: Extremes and Rare Events (1994). ISBN 3-7643-5071-7
Second, revised and extended edition, outside the series (2004): ISBN 3-7643-2416-3

**DMV 22: Knobloch, H.W. / Isidori, A. / Flockerzi, D.**, Topics in Control Theory (1993). ISBN 3-7643-2953-X

**DMV 21: Pohst, M.E.**, Computational Algebraic Number Theory (1993). ISBN 3-7643-2913-0

**DMV 20: Esnault, H. / Viehweg, E.**, Lectures on Vanishing Theorems (1994) (2. printing). ISBN 3-7643-2822-3

**DMV 19: Groeneboom, P. / Wellner, J.A.**, Information Bounds and Nonparametric Maximum Likelihood Estimation (1994) (2nd printing). ISBN 3-7643-2794-4

**DMV 18: Roggenkamp, K.W. / Taylor, M.J.**, Group Rings and Class Groups (1994) (2nd printing). ISBN 3-7643-2734-0

**DMV 17: Ljung, L. / Pflug, G. / Walk, H.**, Stochastic Approximation and Optimization of Random Systems (1992). ISBN 3-7643-2733-2

**DMV 14: Bhattacharya, R. / Denker, M.**, Asymptotic Statistics (1990). ISBN 3-7643-2282-9

**DMV 12: van der Geer, G. / van Lint, J.H.**, Introduction to Coding Theory and Algebraic Geometry (1989). ISBN 3-7643-2230-6

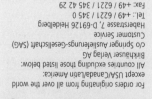

**DMV 12: van der Geer, G. / van Lint, J.H.,** Introduction to Coding Theory and Algebraic Geometry (1989). ISBN 3-7643-2230-6

**DMV 14: Bhattacharya, R. / Denker, M.,** Asymptotic Statistics (1990). ISBN 3-7643-2282-9

**DMV 17: Ljung, L. / Pflug, G. / Walk, H.,** Stochastic Approximation and Optimization of Random Systems (1992). ISBN 3-7643-2733-2

**DMV 18: Roggenkamp, K.W. / Taylor, M.J.,** Group Rings and Class Groups (1994) (2nd printing). ISBN 3-7643-2734-0

**DMV 19: Groeneboom, P. / Wellner, J.A.,** Information Bounds and Nonparametric Maximum Likelihood Estimation (1994) (2nd printing). ISBN 3-7643-2794-4

**DMV 20: Esnault, H. / Viehweg, E.,** Lectures on Vanishing Theorems (1994) (2. printing). ISBN 3-7643-2822-3

**DMV 21: Pohst, M.E.,** Computational Algebraic Number Theory (1993). ISBN 3-7643-2913-0

**DMV 22: Knobloch, H.W. / Isidori, A. / Flockerzi, D.,** Topics in Control Theory (1993). ISBN 3-7643-2953-X

**DMV 23: Falk, M. / Hüsler, J. / Reiss, R.-D.,** Laws of Small Numbers: Extremes and Rare Events (1994). ISBN 3-7643-5071-7
Second, revised and extended edition, outside the series (2004): ISBN 3-7643-2416-3

**DMV 24: Aubry, M.,** Homotopy Theory and Models (1995). ISBN 3-7643-5185-3

**DMV 25: Ballmann, W.,** Lectures on Spaces of Nonpositive Curvature (1995). ISBN 3-7643-5242-6

**DMV 26: Miyaoka, Y. / Peternell, T.,** Geometry of Higher Dimensional Algebraic Varieties (1997). ISBN 3-7643-5490-9

**DMV 27: Knauf, A. / Sinai, Y.G.,** Classical Nonintegrability, Quantum Chaos (1997). ISBN 3-7643-5708-8

**DMV 28: Cercignani, C. / Sattinger, D.,** Scaling Limits and Models in Physical Processes (1998). ISBN 3-7643-5985-4

**DMV 29: Kalai, G. / Ziegler, G.M.** (Eds.), Polytopes — Combinatorics and Computation (2000). ISBN 3-7643-6351-7

**DMV 30: Scholz, E.** (Ed.), Hermann Weyl's Raum—Zeit—Materie and a General Introduction to His Scientific Work (2001). ISBN 3-7643-6476-9

**DMV 31: Huckleberry, A. / Wurzbacher, T.** (Eds.), Infinite Dimensional Kähler Manifolds (2001). ISBN 3-7643-6602-8

**DMV 32: Bolthausen, E. / Sznitman, A.-S.,** Ten Lectures on Random Media (2002). ISBN 3-7643-6703-2

**OWS 33: Kreck, M. / Lück, W.,** The Novikov Conjecture: Geometry and Algebra (2004). ISBN 3-7643-7141-2

852